207

Advances in Polymer Science

AF790170

Advances in Polymer Science
Recently Published and Forthcoming Volumes

Hydrogen Bonded Polymers
Volume Editor: Binder, W.
Vol. 207, 2007

Oligomers · Polymer Composites · Molecular Imprinting
Vol. 206, 2007

Polysaccharides II
Volume Editor: Klemm, D.
Vol. 205, 2006

Neodymium Based Ziegler Catalysts – Fundamental Chemistry
Volume Editor: Nuyken, O.
Vol. 204, 2006

Polymers for Regenerative Medicine
Volume Editor: Werner, C.
Vol. 203, 2006

Peptide Hybrid Polymers
Volume Editors: Klok, H.-A., Schlaad, H.
Vol. 202, 2006

Supramolecular Polymers · Polymeric Betains · Oligomers
Vol. 201, 2006

Ordered Polymeric Nanostructures at Surfaces
Volume Editor: Vancso, G. J., Reiter, G.
Vol. 200, 2006

Emissive Materials · Nanomaterials
Vol. 199, 2006

Surface-Initiated Polymerization II
Volume Editor: Jordan, R.
Vol. 198, 2006

Surface-Initiated Polymerization I
Volume Editor: Jordan, R.
Vol. 197, 2006

Conformation-Dependent Design of Sequences in Copolymers II
Volume Editor: Khokhlov, A. R.
Vol. 196, 2006

Conformation-Dependent Design of Sequences in Copolymers I
Volume Editor: Khokhlov, A. R.
Vol. 195, 2006

Enzyme-Catalyzed Synthesis of Polymers
Volume Editors: Kobayashi, S., Ritter, H., Kaplan, D.
Vol. 194, 2006

Polymer Therapeutics II
Polymers as Drugs, Conjugates and Gene Delivery Systems
Volume Editors: Satchi-Fainaro, R., Duncan, R.
Vol. 193, 2006

Polymer Therapeutics I
Polymers as Drugs, Conjugates and Gene Delivery Systems
Volume Editors: Satchi-Fainaro, R., Duncan, R.
Vol. 192, 2006

Interphases and Mesophases in Polymer Crystallization III
Volume Editor: Allegra, G.
Vol. 191, 2005

Block Copolymers II
Volume Editor: Abetz, V.
Vol. 190, 2005

Block Copolymers I
Volume Editor: Abetz, V.
Vol. 189, 2005

Intrinsic Molecular Mobility and Toughness of Polymers II
Volume Editor: Kausch, H.-H.
Vol. 188, 2005

Hydrogen Bonded Polymers

Volume Editor: Wolfgang Binder

With contributions by
W. H. Binder · L. Bouteiller · G. ten Brinke
O. Ikkala · V. M. Rotello · J. Ruokolainen
S. Srivastava · H. Xu · R. Zirbs

Springer

The series *Advances in Polymer Science* presents critical reviews of the present and future trends in polymer and biopolymer science including chemistry, physical chemistry, physics and material science. It is adressed to all scientists at universities and in industry who wish to keep abreast of advances in the topics covered.
As a rule, contributions are specially commissioned. The editors and publishers will, however, always be pleased to receive suggestions and supplementary information. Papers are accepted for *Advances in Polymer Science* in English.
In references *Advances in Polymer Science* is abbreviated *Adv Polym Sci* and is cited as a journal.

Springer WWW home page: springer.com
Visit the APS content at springerlink.com

ISSN 0065-3195
ISBN 978-3-642-08813-1
e-ISBN 978-3-540-68588-3
DOI 10.1007/978-3-540-68588-3

Springer is a part of Springer Science+Business Media

springer.com

Cover design: WMXDesign GmbH, Heidelberg

Printed on acid-free paper 02/3100 YL – 5 4 3 2 1 0

Volume Editor

Prof. Dr. Wolfgang Binder
Institute of Applied Synthetic Chemistry
Division Macromolecular Chemistry
Vienna University of Technology
Getreidemarkt 9/163/MC
1060 Vienna, Austria
wbinder@mail.zserv.tuwien.ac.at

Editorial Board

Advances in Polymer Science
Also Available Electronically

For all customers who have a standing order to Advances in Polymer Science, we offer the electronic version via SpringerLink free of charge. Please contact your librarian who can receive a password or free access to the full articles by registering at:

springerlink.com

If you do not have a subscription, you can still view the tables of contents of the volumes and the abstract of each article by going to the SpringerLink Homepage, clicking on "Browse by Online Libraries", then "Chemical Sciences", and finally choose Advances in Polymer Science.

You will find information about the

- Editorial Board
- Aims and Scope
- Instructions for Authors
- Sample Contribution

at springer.com using the search function.

Preface

Control of polymeric structure is among the most important endeavours of modern macromolecular science. In particular, tailoring the positioning and strength of intermolecular forces within macromolecules by synthetic methods and thus gaining structural control over the final polymeric materials has become feasible, resulting in the field of supramolecular polymer science. Besides other intermolecular forces, hydrogen bonds are unique intermolecular forces enabling the tuning of material properties via self-assembly processes over a wide range of interaction strength ranging from several kJ mol^{-1} to several tens of kJ mol^{-1}. Central for the formation of these structures are precursor molecules of small molecular weight (usually lower than 10 000), which can assemble in solid or solution to aggregates of defined geometry. Intermolecular hydrogen bonds at defined positions of these building blocks as well as their respective starting geometry and the initial size determine the mode of assembly into supramolecular polymers forming network-, rodlike-, fibrous-, disclike-, helical-, lamellar- and chainlike architectures. In all cases, weak to strong hydrogen-bonding interactions can act as the central structure-directing force for the organization of polymer chains and thus the final materials' properties.

The important contribution of hydrogen bonds to the area of supramolecular polymer chemistry is definitely outstanding, most of all since the potency of hydrogen-bonding systems has been found to be unique in relation to other supramolecular interactions. Thus the high level of structural diversity of many hydrogen-bonding systems as well as their high level of directionality and specificity in recognition-phenomena is unbeaten in supramolecular chemistry. The realization, that their stability can be tuned over a wide range of binding strength is important for tuning the resulting material properties, ranging from elastomeric to thermoplastic and even highly crosslinked duroplastic structures and networks. On the basis of the thermal reversibility, new materials with highly tunable properties can now be prepared, being able to change their mechanical and optoelectronic properties with very small changes of external stimuli. Thus the field of hydrogen-bonded polymers forms the basis for stimuli responsive and adaptable materials of the future. Moreover, the recognition that many aspects of the "bulk"-supramolecular polymer-chemistry can be transferred to binding and recognition events on surfaces is an area still in its infancy. Binding processes of polymers, nanoparticles or other nanosized objects onto (polymeric, quasipolymeric) surfaces by noncovalent interactions already forms a new and strongly expanding area in nanoscience and nanotechnology.

The exploitation of the high specificity of the hydrogen-bonding systems, combined with their dynamic features has opened a new branch in polymer science: dynamic materials with self selection processes. This field, opened up by J. M. Lehn with his "dynamers" is highly prospective for the generation of new materials with properties unachievable with conventional monomers and polymeric materials, relying purely on the covalent bond, instead of the noncovalent, supramolecular interaction.

The present volume on *Hydrogen-Bonded Polymers* provides an overview on these aspects within four main chapters. Different points of view are mirrored, featuring aspects related to (a) classification of hydrogen-bonded polymers according to the nature of the connecting hydrogen bond (by W. H. Binder and R. Zirbs) (b) small-molecule self assembly into hydrogen-bonded polymers (by L. Bouteiller) (c) properties of the resulting materials, with a main focus on the interplay of dynamic properties and polymer-microphases (ten G. Brinke, J. Ruokolainen, O. Ikkala) and (d) nanocomposite materials derived from Hydrogen-bonding elements (H. Xu, S. Srivastava, V. M. Rotello). The varying titles demonstrate that hydrogen-bonded supramolecular polymer chemistry is a highly interdisciplinary research field, where structure, properties and function are closely interrelated to each other.

Still in its infancy, the field of supramolecular polymer chemistry has definitely found its own area and fixed place within the area of macromolecular and polymer chemistry. Although with a certain delay, the recognition of "designed" intermolecular forces as a tool to direct the ordering and function of macromolecules has now been widely acknowledged and respected. The transfer of principles of "organic" supramolecular chemistry is fully accomplished and used with great perfection. Many principles exploited during the past years in this field therefore have already found their application in polymeric material science, and will definitely expand in the near future.

Vienna, February 2007 Wolfgang H. Binder

Contents

Supramolecular Polymers and Networks with Hydrogen Bonds in the Main and Side-Chain
W. H. Binder · R. Zirbs 1

Assembly via Hydrogen Bonds of Low Molar Mass Compounds into Supramolecular Polymers
L. Bouteiller 79

Supramolecular Materials Based on Hydrogen-Bonded Polymers
G. ten Brinke · J. Ruokolainen · O. Ikkala 113

Nanocomposites Based on Hydrogen Bonds
H. Xu · S. Srivastava · V. M. Rotello 179

Author Index Volumes 201–207 199

Subject Index 203

Adv Polym Sci (2007) 207: 1–78
DOI 10.1007/12_2006_109

Published online: 2 December 2006

Supramolecular Polymers and Networks with Hydrogen Bonds in the Main- and Side-Chain

Wolfgang H. Binder (✉) · Ronald Zirbs

Institute of Applied Synthetic Chemistry, Division Macromolecular Chemistry, Vienna University of Technology, Getreidemarkt 9/163/MC, 1060 Vienna, Austria
wbinder@mail.zserv.tuwien.ac.at

1 Introduction 3

2 Hydrogen Bonds 5

3 Main Chain Hydrogen-Bonded Polymers 9
3.1 Monovalent Hydrogen Bonds (Liquid Crystalline Polymers and Polymer-Blends via H-bonding) 10
3.2 Polymers Connected with Bivalent Hydrogen Bonds 12
3.3 Polymers Connected with Trivalent Hydrogen Bonds 18
3.4 Polymers Connected with Quadruple Hydrogen Bonds 28
3.5 Polymers Connected with Multiple Hydrogen Bonds 39
3.6 Applications 51

4 Side-Chain Hydrogen-Bonded Polymers 51

5 Hydrogen Bonds on Surfaces 63

6 Conclusions and Future Outlook 70

References 71

Abstract Control of polymeric structure is among the most important endeavors of modern macromolecular science. In particular, tailoring the positioning and strength of intermolecular forces within macromolecules by synthetic methods and thus gaining structural control over the final polymeric materials has become feasible, resulting in the field of supramolecular polymer science. Besides other intermolecular forces, hydrogen bonds are unique intermolecular forces enabling the tuning of material properties via self-assembly processes over a wide range of interaction strength ranging from several $kJ\,mol^{-1}$ to several tens of $kJ\,mol^{-1}$. The present review provides an overview of hydrogen-bonded polymers, with a focus directed towards the type of hydrogen bond as well as their effect on the final, ordered materials. Thus, the ordering effects of single-, double-, triple-, quadruple and multiple hydrogen bonds are discussed separately. Furthermore, various architectures as well as the use of hydrogen bonds on planar surfaces to assemble quasipolymeric structures are discussed.

Keywords Hydrogen bond · Supramolecular polymer · Surface · Polymeric material · Self assembly

Abbreviations

A	hydrogen-bonding acceptor
AFM	atomic force microscopy
AA-PDMS	diacid telechelic poly(dimethylsiloxane)
A-PDMS	monoacid telechelic poly(dimethylsiloxane)
ATRP	atom transfer radical polymerization
D	hydrogen-bonding donor
DAP	2,6-diamino-pyridine
DBSA	dodecyl benzenesulfonic acid
DMSO	dimethyl sulfoxide
MDI	methylene-4,4′-diisocyanate
NDP	nonadecyl phenol
MMA	methyl methacrylate
NMP	*N*-methyl-morpholine
NMR	nuclear magnetic resonance
PAA	poly(acrylic acid)
PCL	poly(caprolactone)
PDMS	poly(dimethylsiloxane)
PDP	pentadecyl phenol
PEO	poly(ethylene oxide)
PEOx	poly(ethyloxazoline)
PEK	poly(ether ketone)
PI	poly(isoprene)
PIB	poly(isobutylene)
PIPS	polymeric induced phase separation
PMMA	poly(methylmethacrylate)
PPO	poly(propylene oxide)
PS	poly(styrene)
PSSA	poly(styrene-4-sulfonic acid)
PS(OH)	poly(4-hydroxy-phenol)
PVAc	poly(vinylacetate)
PVDAT	poly(vinyldiaminotriazine)
PVP	poly(vinylpyridine)
P4VP	poly(4-vinylpyridine)
ROMP	ring opening metathesis polymerisation
SAM	self assembled monolayer
SAXS	small angle X-ray scattering
SIMS	dynamic secondary ion mass spectrometry
SSL	strong segregation limit
STVPy	styrene-co-4-vinylpyridine
STVPh	styrene-co-4-vinylphenol
STM	scanning tunnelling microscopy
THF	tetrahydrofurane
TDI	toluene-2,4-diisocyanate
WAXS	wide angle X-ray scattering
WSL	weak segregation limit

1 Introduction

Polymeric structure and structure formation has been subjected to strong changes in point-of-view during the past centuries. First, the nineteenth century saw macromolecules viewed as mainly colloids and these were postulated as being aggregates of small organic molecules with the forces involved being "side valence forces or Nebenvalenzen"; however, this view was changed drastically by Herrmann Staudinger [1], who realized that covalent bonds were the building force of macromolecules, which were then identified as the constituent structure of polymers. It was realized that the characteristic properties of polymers were determined by the initial structure of the macromolecules, i.e., the identity of the monomers, the degree of polymerization, the mode of distribution of specific monomers within the chain, the topology of the chains (linear, grafted polymers or dendritic), the stereochemistry (tacticity) and the crosslinking between chains. Furthermore, the ordering of macromolecules was found to be most important in determining the final materials properties by influencing crystallization behavior, phase separation and thus in turn the chemical and mechanical properties of polymers. Most of all, concepts to use weak hydrogen-bonding interactions were found to promote the formation of miscibility and thus the controlled formation of polymer blends [2].

Lehn et al. [3] first introduced the principle of supramolecular polymer chemistry more as an extension of substituting small organic molecules by telechelic polymers than "real" long chain polymer structures. It was suggested that intermolecular forces can be used to assemble small organic molecules into two- and three-dimensional structures reminiscent of linear or crosslinked polymers. Among other forces (such as dipol-dipol interactions, pi-pi stacking and charge-charge interactions) the hydrogen bonds form a central structural building-force to generate chains with low molecular weights as well as dendritic and weakly crosslinked structures in solution. The idea, however, was revolutionary in that it disrupted the concept of conventional polymer chemistry (in the sense of Staudinger), which regarded polymers as covalently bound monomeric units. Therefore, not only the general influence of hydrogen bonds on the bulk-polymeric structure (as in polyamides, polycarbonates, polyesters) or the solution-structure (as in PMMA, PVAc, PVP, PEOx, polyvinylalcohol) is taken into account, moreover the hydrogen bond as a tuned interaction is used for polymer organization. Thus, polymers became a much larger group of materials, which now included even higher molecular weight aggregates with thermally reversible linkages at room temperature or below. A variety of "highly ordered" structures, in particular those on planar surfaces such as Langmuir–Blodgett films, self-assembled monolayers were given the name "two-dimensional polymer".

More highly organized supramolecular polymers were then introduced by Stadler et al. [4] by transforming a linear, noncrosslinked soft polybutadiene chain into a thermoplastic elastomer. The concept relied on the statistical introduction of hydrogen bonds into the polybutadiene thus creating defined noncovalent crosslinking sites between the polymer chains. It was found that a hydrogen binding donor-acceptor unit for each 50 butadiene-units (i.e.: at approx. 2 mol %) is sufficient to drastically change the thermal and mechanical properties of the material, thus demonstrating that a couple of weak bonds on the side of a polymer chain can be very efficient in determining the final materials properties.

From this the concept of supramolecular polymers has evolved [5, 6]. Central for the formation of these structures are precursor molecules of small molecular weight (usually lower than 10 000), which can assemble in solid or solution to aggregates of defined geometry. Intermolecular hydrogen bonds at defined positions of these building blocks as well as their respective

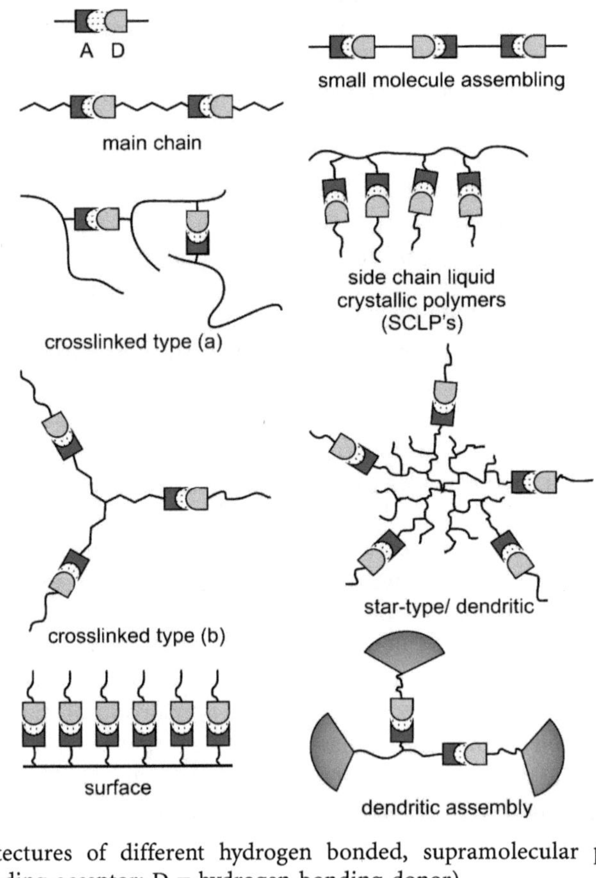

Fig. 1 Architectures of different hydrogen bonded, supramolecular polymers (A = hydrogen-bonding acceptor; D = hydrogen-bonding donor)

starting geometry and the initial size determine the mode of assembly into supramolecular polymers. Ordering can form network-, rodlike-, fibrous-, disclike-, helical-, lamellar- and chainlike structures as depicted in Fig. 1. In all cases, weak-to-strong hydrogen-bonding interactions can act as the central structure-directing force for the organization of polymer chains and thus the final materials properties.

The present review focuses on the formation of macromolecular structure via hydrogen bonds based upon supramolecular concepts, with a focus on the literature published between 2000 and 2006. There are a number of recent reviews relating to the topic of hydrogen-bonded supramolecular polymers, mostly with a focus on specific hydrogen-bonding systems [7–13]. The last review on hydrogen-bonded polymers, related to a broader view on hydrogen-bonding systems by Meijer et al. in 2001 [14] is therefore taken as a starting point for the newer literature. Thus, the focus is directed at designed interactions, deriving from either a functionalized monomer, or a polymeric endgroup, including hydrogen-bonding systems only. The association of molecules with a small molecular weight (below 1000 Dalton) will not be covered, since the work by L. Bouteiller in this series deals with this very aspect. Sect. 2 will give a short overview of hydrogen bonds, a compilation on their strength as well as those used in supramolecular polymer chemistry. Sect. 3 will deal with polymers bearing hydrogen bonds in their main chain, ordered according to the number of hydrogen bonds involved. Sect. 4 deals with polymers bearing hydrogen bonds in their side-chain. Sect. 5 focuses on the use of hydrogen bonds on surfaces, to bind polymers or generate quasipolymeric structures.

2
Hydrogen Bonds

Hydrogen bonds as intermolecular forces have been reviewed intensely in books [15] and reviews [16]. In principle three different classes of hydrogen-bonding systems are discriminated (Fig. 2) (a) strong hydrogen bonds, (b) medium or weak hydrogen bonds, and (c) nonclassical hydrogen bonds. Jeffrey and Sanger classify strong hydrogen bonds as those where two center bonds (such as $F-H\cdots F^-$; $O-H\cdots O^-$; $O^+-H\cdots O$ bonds) are involved, which display short distances, a strongly directional nature and association energies higher than $\sim 40\,kJ\,mol^{-1}$. Medium and weak hydrogen bonds are classified by a $D-H\cdots A$ structure, where directionality is partially lost, and the bond energies are between $20–40\,kJ\,mol^{-1}$. Usually, the residue A is strongly electronegative, whereas the residue D may be electronegative or even a carbon atom. Nonclassical (also termed unconventional) hydrogen bonds [17] involve the interaction of $D-H$ with $A=\pi$-systems as well as transition metals (interaction directly with the metal or via the metal hydride) or boron hy-

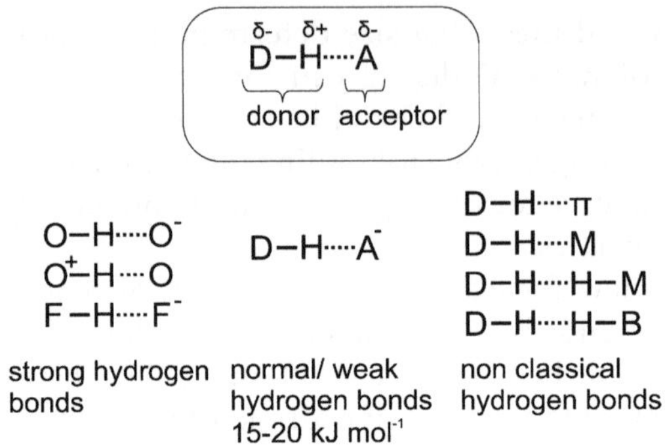

Fig. 2 Overview on various hydrogen bonds ranging from strong to nonclassical hydrogen bonds (D = hydrogen-bonding donor; A = hydrogen-bonding acceptor; M = metal; B = boron)

drides. In general, the strength of each individual hydrogen bond is strongly dependent on solvent effects, most of all polar and protic solvents. It has been demonstrated quite often, that the addition of a polar solvent significantly lowers the hydrogen bond over many orders of magnitude. Therefore, the supramolecular chemistry of hydrogen-bonded polymers is mostly done in aprotic and nonpolar solvents such as linear and cyclic alkanes, toluene, dichloromethane and chloroform.

The main parameter determining the strength of a hydrogen-bonding system is the number of individual bonds involved. Thus, as a rule of the thumb, more hydrogen bonds imply a stronger binding interaction, with the ideal value of about 7.4 kJ mol^{-1}/hydrogen bond. Figures 3–5 list the most prominent hydrogen-bonding systems used in supramolecular chemistry of polymers. Starting from those with only one hydrogen-bonding interaction (Fig. 3), two-centered hydrogen bonds (Fig. 4), three- (Fig. 5), four- and multiple hydrogen-bonding interactions (Fig. 6) are listed. In contrast to other

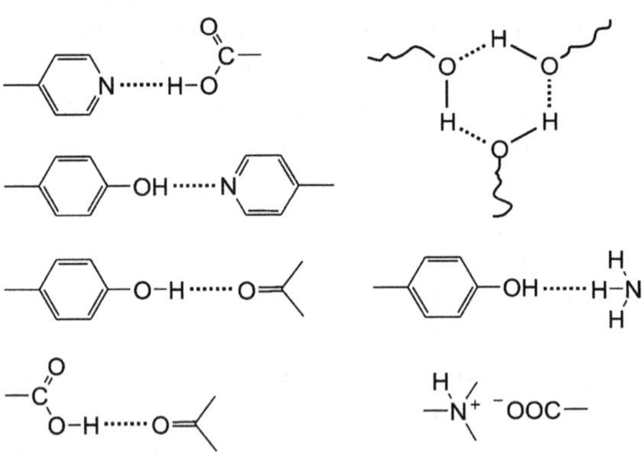

Fig. 3 Molecular structures of single hydrogen bonds

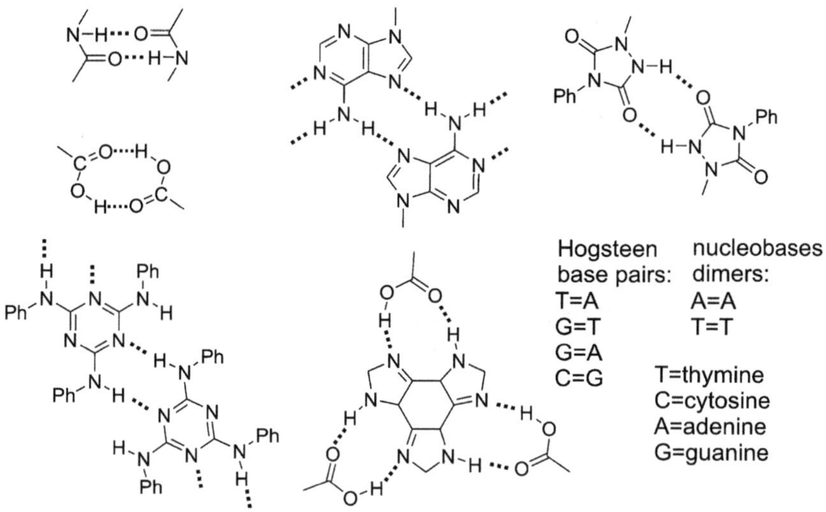

Fig. 4 Molecular structure of two-centered hydrogen bonds (A = adenine; T = thymine; G = guanine; C = cytosine)

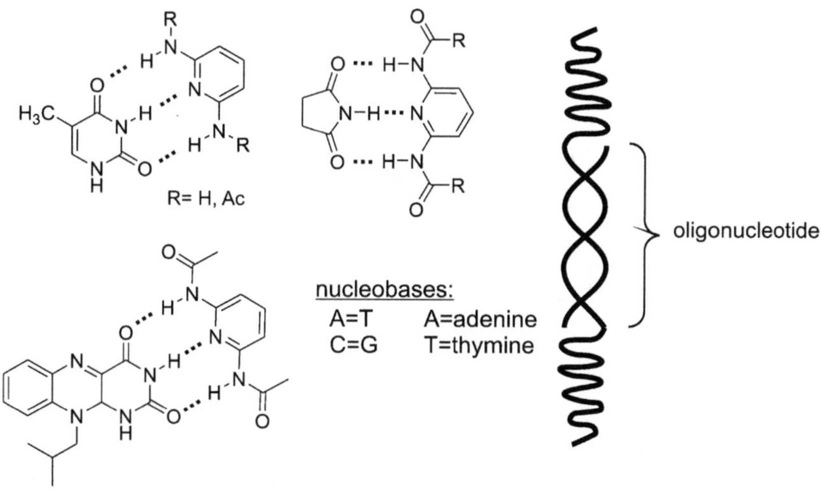

Fig. 5 Examples of triple-hydrogen-bonding interactions

supramolecular interactions (such as metal-metal complexes) the strength of the hydrogen-bonding systems are tunable over a wide range of association energies and—often—display a unique host-guest complementarity, thus excluding self-association phenomena, leading to a highly controllable level of intermolecular ordering.

Besides the number of hydrogen bonds and solvents effects, other effects can strongly determine the strength of the hydrogen-boding systems: (a) sec-

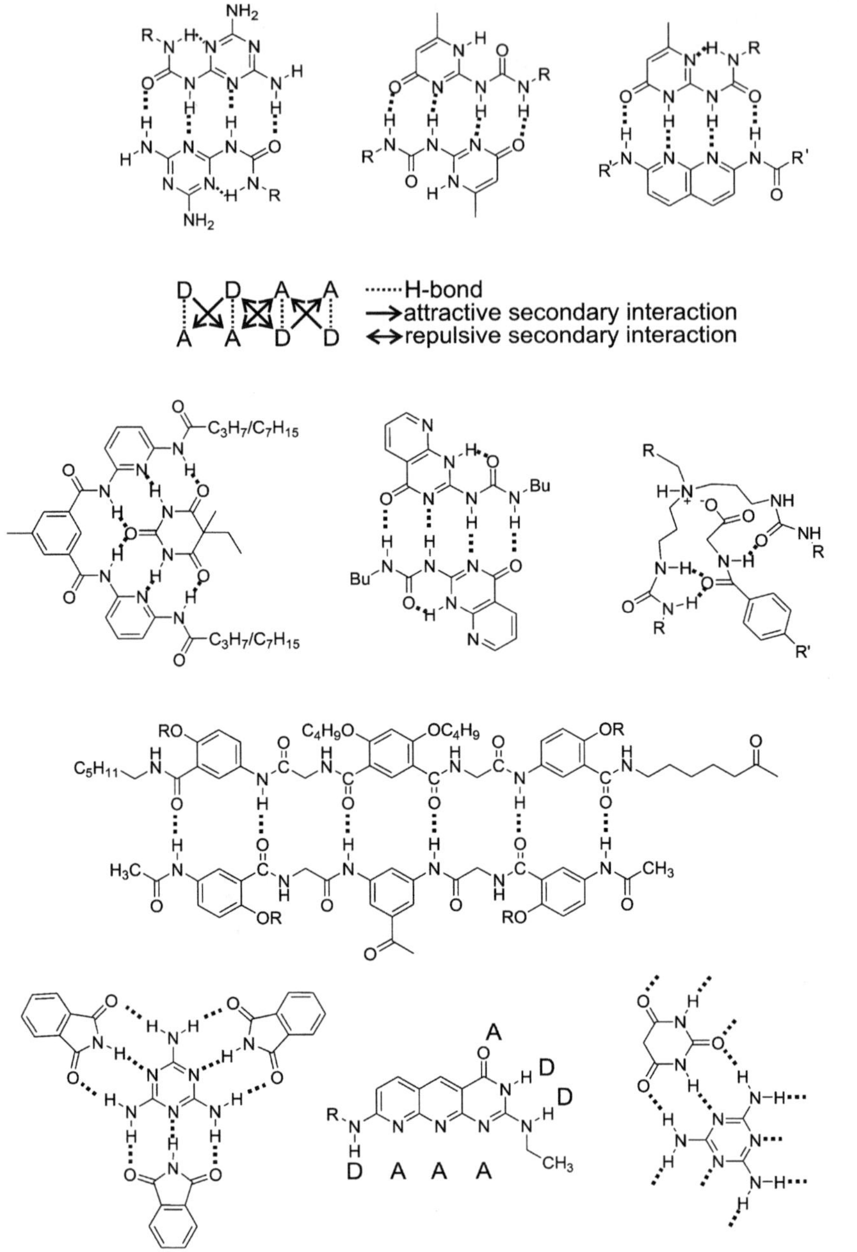

Fig. 6 Examples of quadruple and multiple hydrogen-bonding interactions

ondary interactions can strongly increase or decrease the strength of multiple hydrogen-bonding systems. Thus, Schneider et al. [18] have developed an incremental system to describe secondary interactions by adding or subtracting the secondary interaction energy of approx. $\pm 2.9\,\mathrm{kJ\,mol^{-1}}$. (b) Electronic effects can determine the strength of similar hydrogen-bonding systems, most of all in aromatic systems [19]. (c) Preorganizational and tautomeric effects have a strong influence on hydrogen-bonding strength and interactions. Most prominently, these effects were demonstrated by Meijer et al. [20, 21] as well as recent investigations using other multiple hydrogen bonds [22]. Thus, the preorganization of complex hydrogen-bonding codes via internal fixation to prevent rotational freedom has been found extremely useful in this respect [22].

An important aspect in the field of hydrogen-bonded polymers concerns the determination method of the binding energy, association constant as well as the underlying dynamics of the bonds. The often underlying complex equilibria are investigated using solution NMR spectroscopy [23, 24] and often complex mathematical fitting procedures. Isothermal titration calorimetry [25, 26] is a highly efficient tool to determine the association energy and extracting free enthalpy, entropy and enthalpy of the binding process. IR spectroscopy in solution and the solid state, as well as solid-state NMR spectroscopy have been found to be powerful tools to investigate the nature and dynamics of the hydrogen bonds [27, 28]. The time scale of breaking/reforming hydrogen bonds has been extensively investigated [29].

3 Main Chain Hydrogen-Bonded Polymers

The most simple structures derived from hydrogen-bonded polymers are those where the hydrogen bond is located directly in the main chain. The significant factor in this architecture for the formation of polymeric structures is therefore the strength of the hydrogen-bonding systems, acting as a "glue" for the assembly. Thus, three parameters are important: (a) the nature of the hydrogen bond (i.e.: if the hydrogen is homocomplementary or heterocomplementary), (b) the absolute number of units bound together by hydrogen bonds (i.e.: chemical structure, length, polydispersity and—most of all—miscibility). Relevant aspects in this respect are the correlation between association (mediated by the strength of the H-bonding system) and the molecular mobility of the polymer chains and—eventually—their microphase separation. Some computational publications have dealt with this aspect [30, 31]. The associative process with a "virtual" molecular weight of the associate has been investigated and it was found that considerable strengths of hydrogen bonds are necessary to achieve high molecular weight associates [11]. Thus, monovalent hydrogen-bonding systems may lead to liquid crystallinity, given that additional ordering interactions (such as pi–pi-stacking; hydrophobic- or dipolar interactions) are included into the systems. Therefore, the presence of hydrogen-bonding induces more stable liquid crystalline structures. Strong effects of weak, single hydrogen bonds are only observed in the solid state, where multiple bonds may induce miscibility effects in polymeric blends. With bivalent interactions, the formation of polymeric systems from monomeric units is already feasible. However most of all if multivalent networks can be generated, whereas the effects usually are low in solution, but more pronounced in the solid state. Thus, large modification in the (melt-)rheological properties, as well as in the generation of elastomeric behavior via reversible crosslinking by the hydrogen bonds can be observed. Most importantly, a strong dependency of the materials properties on temperature is observed. These effects are more pronounced if stronger, usually trivalent hydrogen bonds are used. However, since hydrogen bonds can form also higher ordered networks in the solid state, the discrimination of effects stemming from a singular interaction of a multiple interaction is already difficult. In addition to the aforementioned elastomeric and rheological properties, miscibility effects related to microphase separation become more pronounced. Especially if the formation of hydrogen-bonded networks is induced, can the materials properties be altered by only a few hydrogen bonds per several thousand monomeric units. It was for these effects that quadruple and higher-order hydrogen bonds have been developed, most of all by Hamilton et al. [32] and Meijer et al. [21, 33] who used hydrogen bonds with association constants $> 10^5\ M^{-1}$. Using these bonds, the changes in a materials properties and ordering are usually dramatic, changing the starting polymeric or oligomeric materials significantly in their properties. Therefore, in the following sections, the monovalent, bivalent, trivalent, quadruple and multiple hydrogen-bonding systems within polymers are discussed.

3.1 Monovalent Hydrogen Bonds (Liquid Crystalline Polymers and Polymer-Blends via H-bonding)

The topic of monomeric, oligomeric or polymeric building blocks held together with only a single hydrogen bond is strongly related to supramolecular main chain liquid crystalline polymers. A variety of supramolecular forces has been used to arrange molecules of small molecular weight into liquid crystalline materials, most of all using metal-metal complexes, pi–pi interactions, ionic bonds and dipol-dipol interactions, mainly with the demand to increase the dynamic nature of the resulting materials. According to Fig. 7 [34] the structures can be arranged in a dimeric A-B type, with the hydrogen-bonding unit combining the two residues in a reversible manner. Other architectures include the trimeric A-B-A case as well as all higher structures with side chains, combined types and network types. The topic of hydrogen-bonded liquid crystalline materials was reviewed first by Kato et al. in 2001 [35] and later in 2006 [36] and also by Zimmermann et al. [37]. Ad-

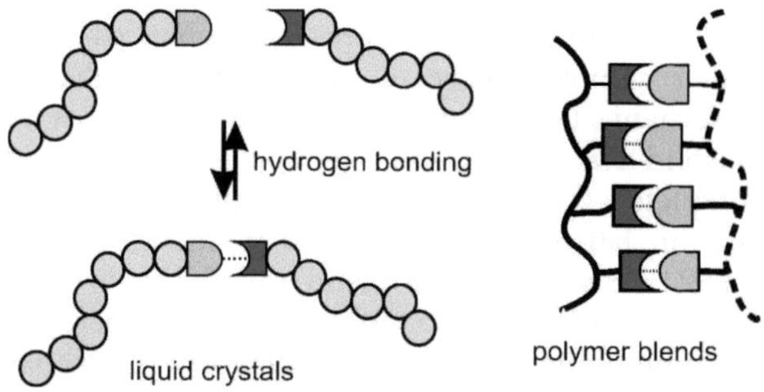

Fig. 7 Formation of liquid crystalline materials or polymer blends by single hydrogen bonds. Redrawn according to reference [34]

ditionally, ten Brinke et al. report their review in this volume. The interested reader is therefore referred to these recent reviews, since all the mentioned topics are covered.

Another important aspect of single hydrogen bonds concerns interpolymer complexation by multiple, but weak hydrogen bonds between immiscible polymers. Weak hydrogen bonds may be those between i.e.: phenolic OH/carbonyls; C – H bonds/ether-bonds in PEO; – COOH/HO – CH_2 moieties; amidic bonds/carbonyls; – CH_2 – OH/carbonyls. A strong miscibility enhancement has been observed, if the number of interactions is large, although the individual interactions may be quite weak. Thus, even weak hydrogen bonds such as those between PMMA/PEO but also stronger bonds [i.e.: those between PS(OH)/PMMA, PS(OH)/PCL; poly(acetoxystyrene)/poly(ethyleneoxide), poly(acrylic acid)/poly(ethyleneoxide)] can be used to induce enhanced miscibility. This topic was reviewed recently by Jiang et al. [38] and will be only roughly updated here. Interested readers are referred to a recent review on micelle formation using the same principle by Jiang et al. [39, 40]. Briefly, the topic of blends in the solid state will be summarized here.

Ming et al. [41–43] studied the surface segregation in polymer blends with increasing hydrogen-bonding interactions by XPS and TOF-SIMS methods. The polymers used were blends consisting of poly(styrene-co-4-vinylphenol)/poly(styrene-co-4-vinylpyridine) (STVPh/STVPy). Up to about 20 mol % of the STVPh, the surface enrichment of the lower surface component (STVPh) was observed. At higher concentrations, the surface enrichment of STVPh is hardly detectable, presumably due to interpolymer complexation.

The pure bulk-phase behavior was studied on blends consisting of poly(acrylic acid)/hydroxypropyl cellulose [44], poly(2-vinylpyridine)/end-sufonic acid poly(styrene) [45], and poly(acrylic acid)/poly(*N*,*N*-dimethylacrylamide) [46]. Goh et al. [47, 48] and others [49] have studied an interesting example of [60]fullerenated poly(2-hydroxyethyl methacrylate) with poly(1-vinylimidazole) or poly(4-vinylpyridine) or poly(styrene-co-4-

vinylpyridine). The addition of 2.6 wt% C60 enhanced the tendency to form interpolymer complexes due to its strong hydrophobic interaction. The dynamic mechanical behavior was studied, as well as the blend homogeneity via solid state NMR spectroscopy [50]. They found that the mixing was complete even at a scale of 1–3 nm, as verified by comparing the proton spin lattice relaxation times in the rotating frame. Other studies using ESR methods on the blend miscibility of poly(styrene-co-methacrylic acid)/poly(butylmethacrylate) were performed by Qiu et al. [51], revealing the maximum miscibility at about 29 mol% of carboxylic acid. Another method probing the heterogeneity in blends consisting of poly(vinyl ethylether)/poly(styrene-co-vinylphenol) relied on dielectric relaxation spectroscopy [52], revealing a larger dynamic heterogeneity in the copolymers in relation to the blends with the pure homopolymers. Jiang et al. [53] compared the ionic interaction with the hydrogen-bonding interaction in mediating polymer miscibility. They concluded, that the hydrogen-bonding interaction mediates a better miscibility effect on the polymeric microphases than the corresponding ionic interaction. Goh et al. [54] have investigated the miscibility of blends consisting of poly(acryloyl-*N*′-phenylpiperazine) with acidic polymers such as poly(acrylic acid), poly(-vinylphenol), poly(styrenesulfonic acid) and poly(vinylphosphonic acid). FT-IR and XPS indicated a purely hydrogen-bonding interaction in the case of PVPh and PAA, whereas a combination of hydrogen bonding and ionic interaction was observed with PSSA and PVPA. Thus, the strength of the acid is a significant factor in the nature of the interaction. Other systems concerning the nature of the hydrogen-bonding interactions in blends have been investigated recently [55–57], as well as carboxyl-containing PDMS/poly(1-vinyl-imidazole) [58], poly(*N*-acryloylmorpholine)/poly(vinylphenol) [59] and in resin materials by Ni et al. [60].

3.2 Polymers Connected with Bivalent Hydrogen Bonds

Bivalent hydrogen bonds are usually stronger than their monovalent counterparts, and therefore the impact of a few hydrogen bonds on the resulting material properties is more pronounced.

Stadler et al. [4, 61] introduced the uradiazole moiety to generate reversible crosslinks within a polymeric material (Fig. 8) thus transforming a linear, noncrosslinked soft polybutadiene chain into a thermoplastic elastomer. The concept relied on the statistical introduction of hydrogen bonds into the polybutadiene by use of an ene-type reaction, thus creating noncovalent crosslinking sites between the polymer chains. It was found that a hydrogen binding donor-acceptor unit for each 50 butadiene units (i.e.: at 2 mol%) is sufficient to drastically change the thermal and mechanical properties of the material, thus demonstrating that a couple of weak bonds on the side of a polymer chain can be a very efficient crosslinking structure.

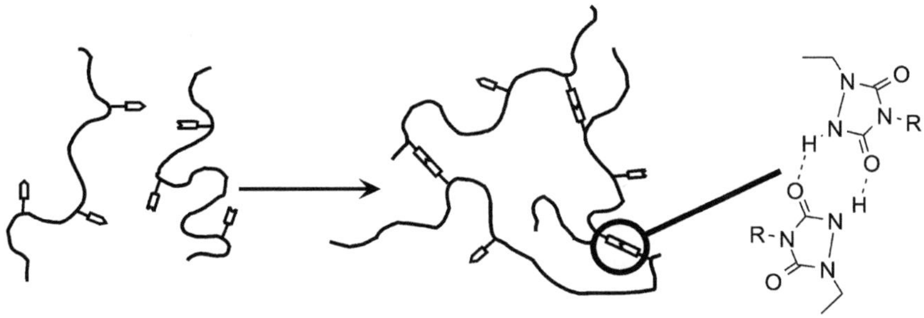

Fig. 8 Ordering of flexible polymer chains by uradiazole-hydrogen bonds into networks

The resulting thermoreversible network displays elastic and thermoelastic properties, due to the formation of domains within the polymer consisting of aggregated domains of associated polymer-chains disperged within the residual polymeric matrix. Rheological thickening effects are observed in solution, where the increased crosslinking density leads to non-Newtonian behavior. The reversibility of the hydrogen bond at higher temperatures or at increased stress leads to tuneable behavior within small temperature intervals. The concept has been applied to many polymers such as poly(butadienes) [62–65], poly(isobutylenes) [66] and poly(siloxanes) [62, 67–69].

Important studies have been done using the bivalent association of the amidic bond. Thus, a multivalent amidic interaction has been used to assemble oligoaromatic building blocks into sheet-type order [70]. The self assembly is achieved in *N*-methylpyrrolidine solution, furnishing the assembled lyotropic liquid crystalline material (Fig. 9). The material bears silanol endgroups and can thus condense into polysiloxanes upon acidification. Surprisingly, this process works under action of sulfuric acid in the aforementioned NMP-solution. The condensation process was followed by ^{29}Si-NMR spectroscopy, proving the formation of a ladder-type structure. Further thermal investigations revealed a siloxane polymer with a T_g of $\sim 280\,^\circ$C, in contrast to uncrosslinked poly(dimethylsiloxane) displaying a T_g of $-100\,^\circ$C.

The amidic bonds within amino acids can be also used to effect the organization of polymers into superstructures (Fig. 10). Thus, the formation of artificial helices on the basis of assembling polymers has been described by use of poly(acetylenes) bearing pendant L-valine side-chains. [71, 72] Two effects are important for the association of these ladder-type polymers into double-stranded helices: (a) the reduction of conformational freedom by the poly(acetylene) chain with respect to a conventional alkyl-chain and (b) the selective association of the L-valine residues by specific hydrogen bonding. An AFM image of the associates on a flat surface demonstrates the presence of a string-pearl structure reminiscent of natural DNA.

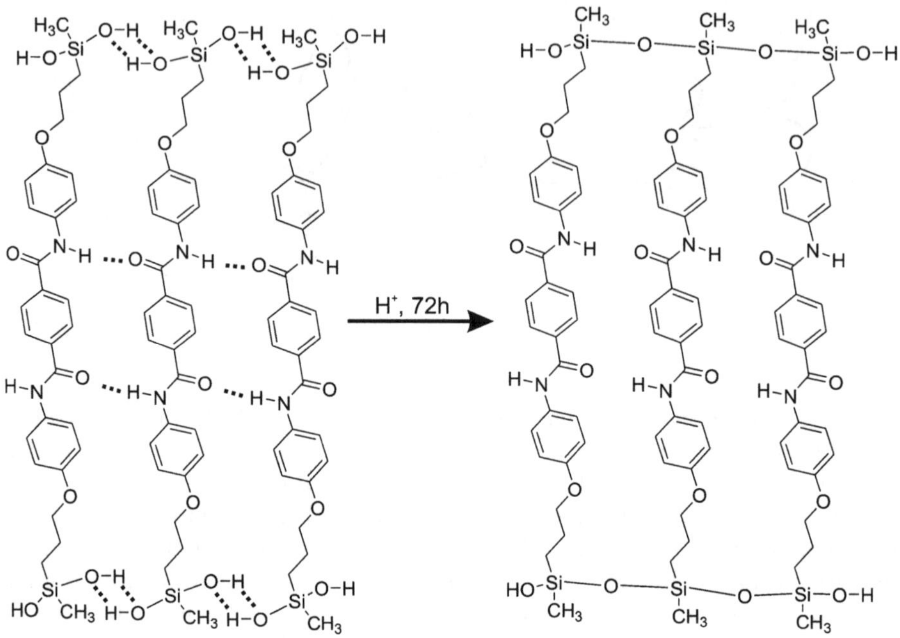

Fig. 9 Self assembly via amidic hydrogen bonds and fixation of the assembly via a sol/gel process

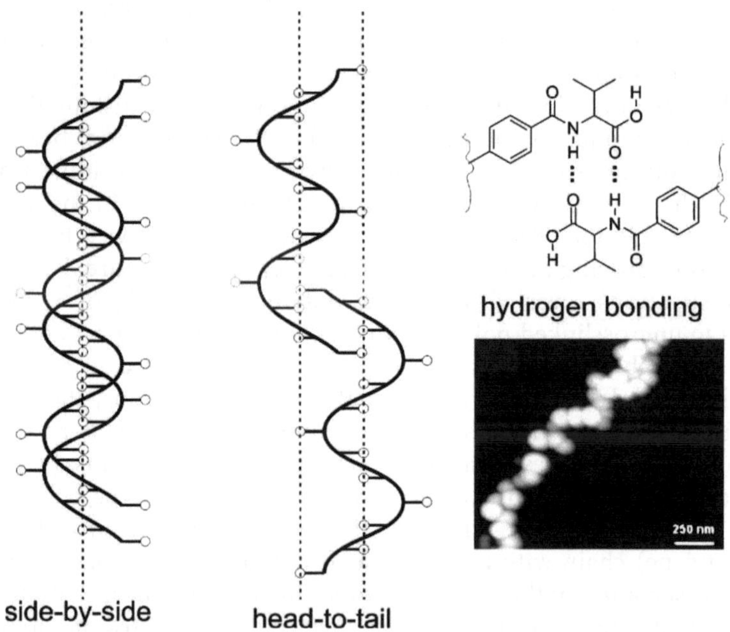

Fig. 10 Self assembly of poly (acetylenes) bearing pendant L-valine moieties into helical superstructures. Reprinted with permission from [71]

Another example of the use of simple amidic bonds for self assembly has been related to dendrimers. Thus, Kim et al. [73] have prepared a dendronic structure displaying amidic bonds at the outer core and a polymerizable 1,3-diine moiety at the inner core (Fig. 11). The first- and second-generation dendrimers show a strongly solvent-dependent association; in chloroform/*n*-hexane a gel is formed, whereas the higher-generation dendron forms a round-shaped aggregate in water. UV-irradiation leads to the stabilization of the gels by topochemical polymerization and thus the formation of lamellar or hexagonal phases as well as micelles.

Carboxylic acids are highly efficient hydrogen-bonding systems for the formation of ordered supramolecular aggregates. Sleiman et al. [74] investigated

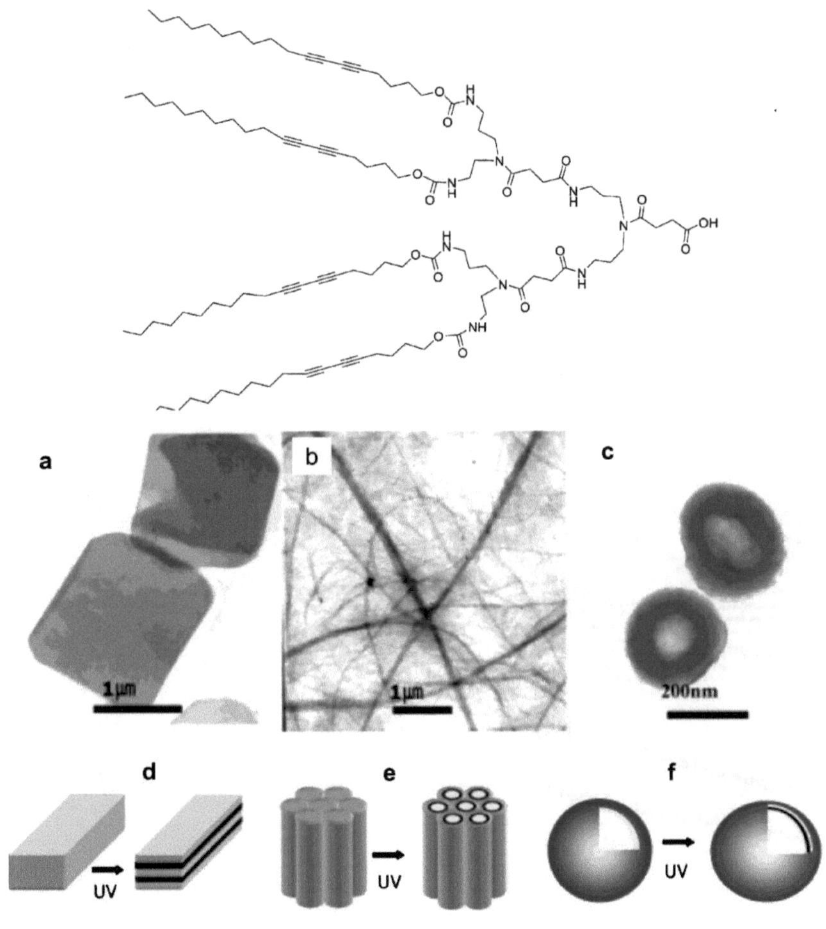

Fig. 11 Formation of aggregates from dendritic-assembly: **a** aggregates in ($CHCl_3$/*n*-hexane = 5/5); **b** in toluene; **c** in water; (**d**, **e**, **f**) stabilization of the aggregates via UV-irradiation. Reprinted with permission from [73]

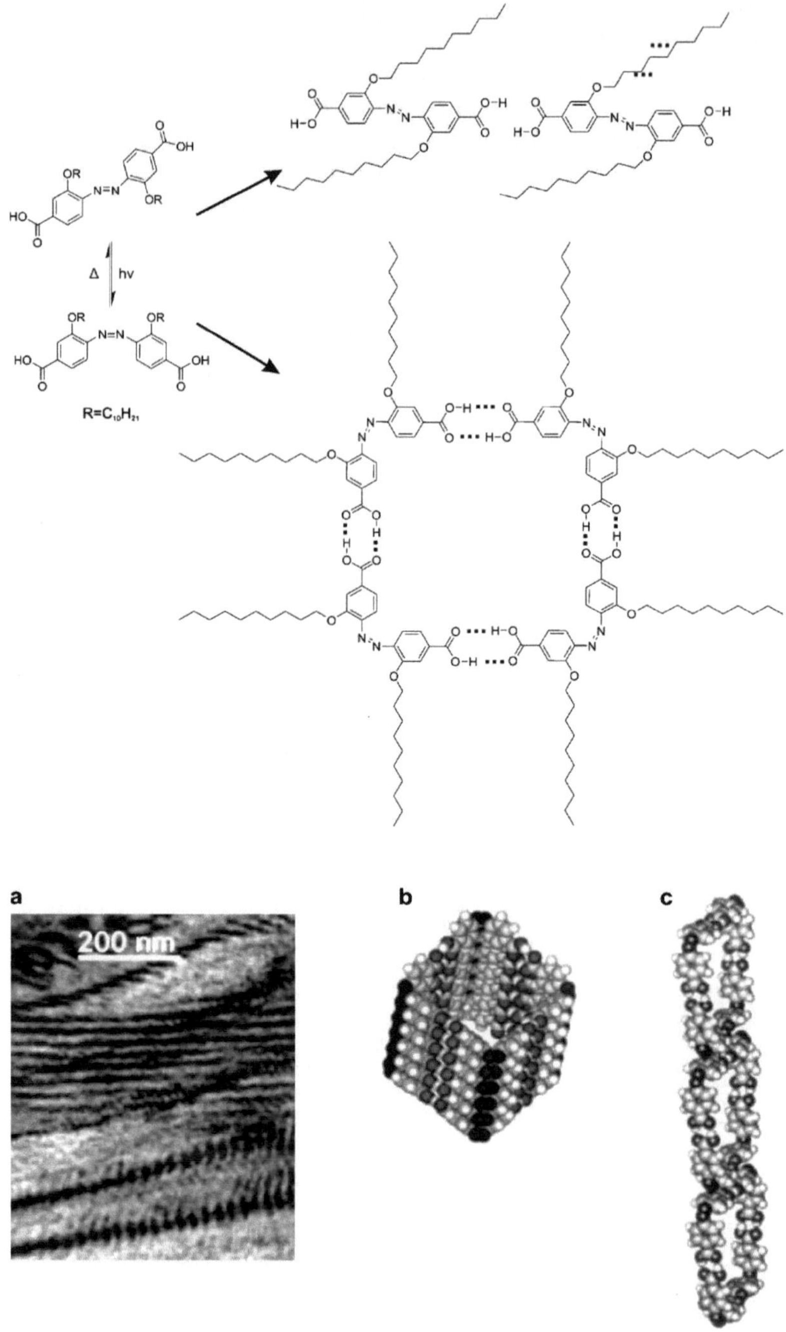

Fig. 12 Assembly of azobenzenes via carboxylic acid dimers. **a** TEM-micrograph of elongated aggregates; (**b**, **c**) models for rod- and fiber-type structures. Reprinted with permission from [74]

the association of azobenzene-linked diacids. Depending on the *cis/trans* position of the azobenzene bond, either linear (*trans*) or cyclic associates (*cis*) were observed (Fig. 12). TEM-studies revealed the fibers generated from the *cis*-form upon deposition on carbon grids. Similar studies by Zimmermann et al. [75] have investigated conventional benzoic acid derivatives with attached dendrons.

The use of simple benzoic acids for the self assembly of polymers has been investigated by Bouteiller et al. [76–78] in a series of papers (Fig. 13). They used telechelic poly(dimethylsiloxane)s with a molecular weight of $\sim 1500\,\mathrm{g\,mol^{-1}}$ capped with either benzoic acid endgroups (AA-PDMS; A-PDMS) or their corresponding esters, the former being able, the latter not able to interact via hydrogen bonds (EE-PDMS). As demonstrated via the relaxation times by ^{13}C-NMR the associates strongly display polymeric behavior revealing the reduced chain mobility due to the association via the hydrogen bonds. The virtual molecular weight derived from these relaxation studies is in the order $200\,000\,\mathrm{g\,mol^{-1}}$, thus "virtually" ordering about 100 telechelics into a polymer.

A rather selective double bond relying on the self aggregation of adenine has been used by Sleiman et al. [79] to organize homo- and block copolymers derived from oxy-norbornenes (Fig. 14). Thus, the homo- and block copolymers prepared by ROMP using a ruthenium-alkylidene catalyst (Grubbs catalyst) and a succinimide complex as a supramolecular protecting group showed—quite unexpectedly—aggregation into rods in a hexane solu-

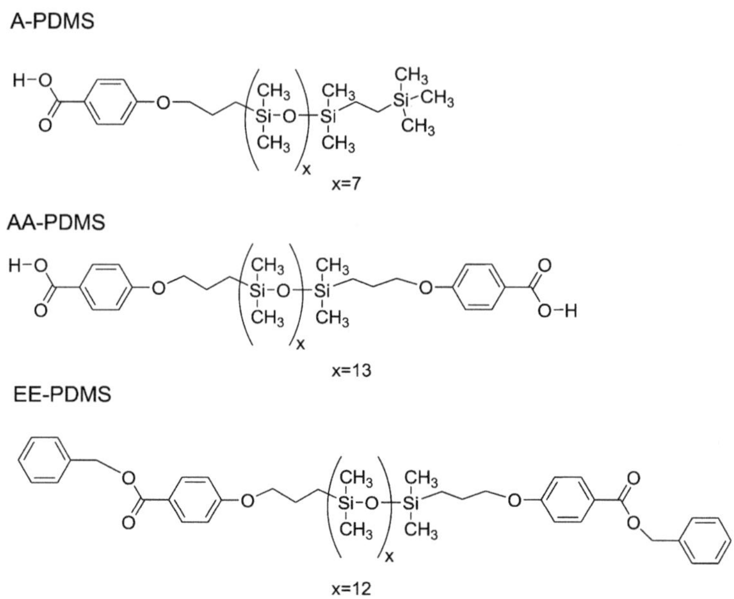

Fig. 13 Assembly of telechelic poly(dimethylsiloxanes) by carboxylic acid dimers

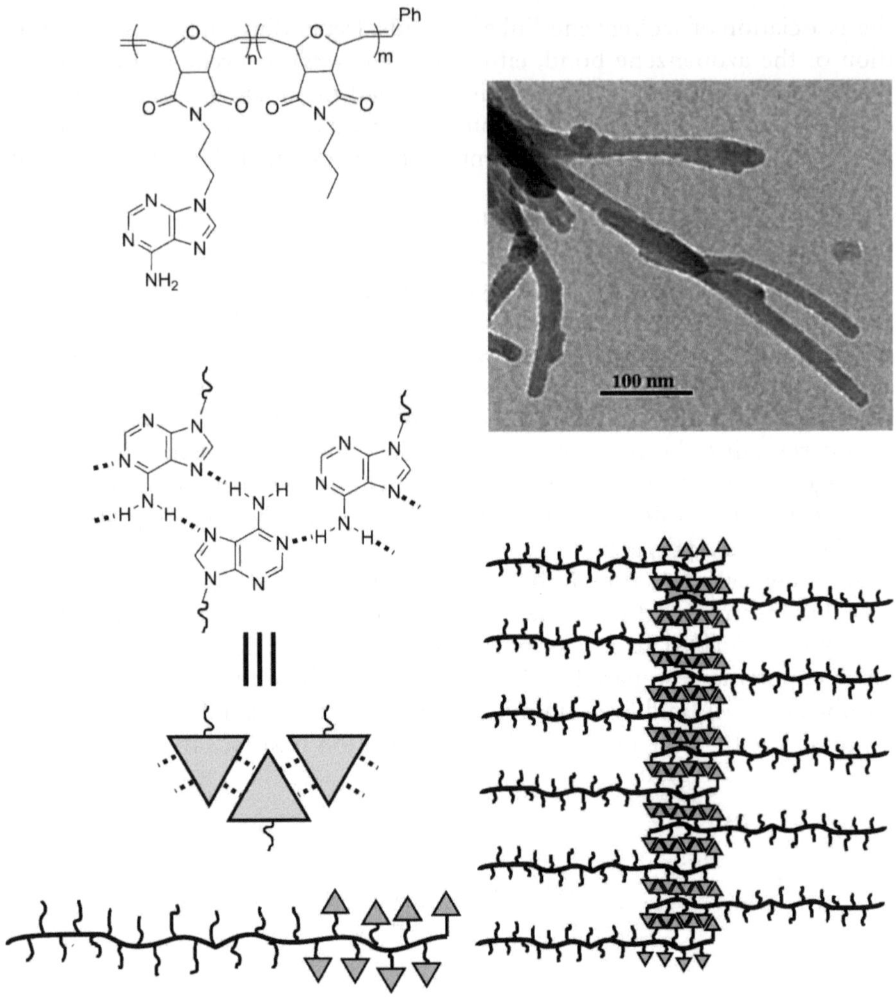

Fig. 14 Assembly of adenine-bearing polymers by self-complementarity. Reprinted with permission from [79]

tion. The self-aggregation seems to be induced by the network formed from the hydrogen bonds, as demonstrated by WAXS measurements.

3.3 Polymers Connected with Trivalent Hydrogen Bonds

Trivalent hydrogen-bonding systems have been used very extensively for guiding and influencing the structuring of polymers. As mentioned in Sect. 1, Fig. 5, the most important triple-hydrogen bonds derive from 2,6-diaminopyridines, 2,6-diamino-1,3,5-triazines and their complexes with flavine- and thymine/uracil, as well as succinimide derivatives. Of course equally important are the nucleobase interactions (adenine/thymine; cytosine/guanine) similar to DNA- and RNA molecules. Although the adenine/thymine interactions are just a bivalent interaction, it is treated most often in combination or comparison with the other bonds. Therefore, the bonds related to nucleobase interactions are all treated in this section. The topic of nucleobase-associated structures has been reviewed recently by Rowan et al. [80]. The important aspect of nucleobases lies in the fact, that not only the well-known Watson–Crick base pairing is possible, but also the less well-known Hogsteen base pairing. Thus, there is a stronger versatility of the bonds as well as the possibility to tune interactions by well-known slight modifications of the heterocyclic structures.

Lehn et al. [81] were among the first to recognize and exploit the importance of the 2,6-diamino-pyridine/uracile interaction (Fig. 15). This is a typically trivalent hydrogen bond, which has been used to assemble chiral moieties derived from tartaric acid. Columnar superstructures were formed, displaying liquid crystalline properties of the resulting associates ranging from temperatures below RT up to about 200 °C, similar to main chain supramolecular liquid crystals. With solvents such as THF, dioxane, $CHCl_3$, gels are formed, presumably due to fiber formation. A similar approach with a trivalent hydrogen bond was reported in 1995 [82] (Fig. 15b).

An important contribution towards the effect of triple hydrogen bonds on polymer behavior was demonstrated by Meijer et al. [83] in early 1995.

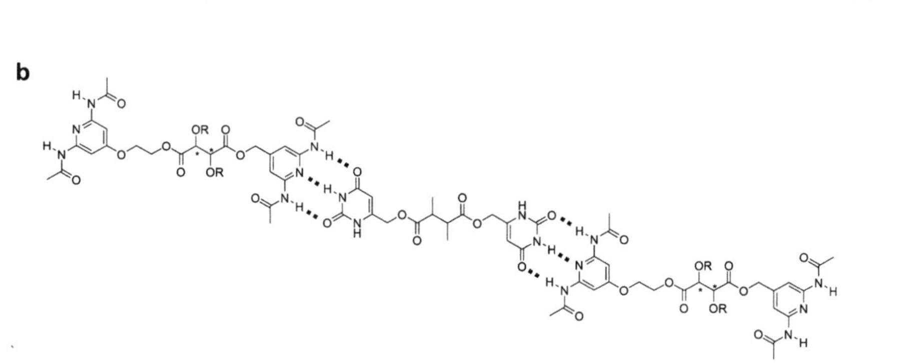

Fig. 15 Fiber formation via self assembly of building blocks bearing triple bonds

Here, an alternating copolymer consisting of PS-maleimide was blended with melamine up to 2.6 mol-equivalents. As expected, a superstructure was formed, indicating a 1 : 3 organization via the hydrogen-bonding moieties. The blends, that were prepared from solution, showed good mixing at the microscale as indicated by a single T_g.

The interaction of multiple sites between poly(vinyldiaminotriazine) (PV-DAT) with small molecules interacting via matching hydrogen bonding has been studied for quite some time [84] (Fig. 16). The relevant question in this endeavor was the interaction with various pyrimidine- and purine derivatives in aqueous solution. Thus, the binding of cytosine, 3-methyluracile, pyrimidine, xanthine, theobromine, adenine, caffeine, guanine and purine was studied. Binding in water was possible, since the PVDAT provided a quasi-hydrophobic micro-environment for the binding process. When aromatic moieties were introduced near the hydrogen-bonding entities, the binding process was enhanced. Similar results have been found for the binding of nucleoside-5′-monophosphates [85].

Hierarchical ordering into mesoscopic superstructures of perylene derivatives with strongly hydrophobic diamino-dialkyl-triazines was reported by Würthner et al. [86] (Fig. 17). Well-defined mesoscopic structures with high photostability and high fluorescence quantum yield were formed. Critical was the exploitation of several interactions, most of all hydrogen-bonding interactions together with pi–pi stacking and hydrophobic side-chains, yielding the superstructures at very low concentrations (10^{-6} mol/L). Cylindrical strands formed with diameters of roughly 200–300 nm in apolar solvents such as methylcyclohexane.

Long et al. [87] have studied the association of PS-polymers bearing adenine- and thymine endgroups, prepared via Michael-type addition. The association was followed by ^{1}H-NMR spectroscopy, relying on the temperature dependency of the imide-protons and revealed strong effects in solution. The formation of 1 : 1 complexes was proven and extended to other polymers such as poly(acrylates) and their respective melt viscosities [88]. Similar results have been reported for PS-b-PI [89] and polyisoprenes [90]. In

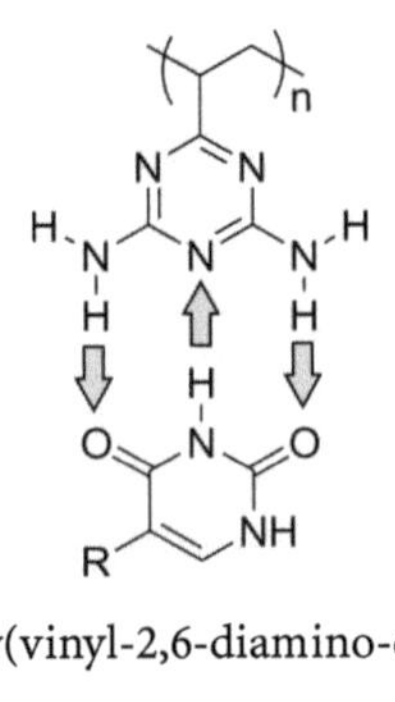

Fig. 16 Interaction between poly(vinyl-2,6-diamino-diaminotriazine) with pyrimidine derivatives

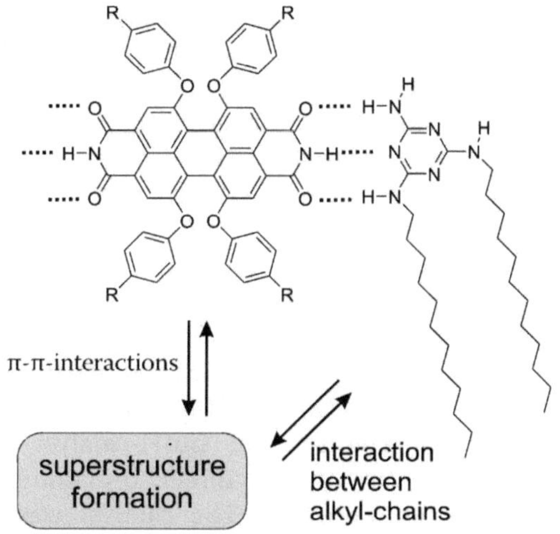

Fig. 17 Formation of cylindrical strands from perylene/2,4,6-triaminotriazine aggregates. Redrawn according to [86]

the latter systems strong rubber-like properties were observed using maleic-anhydride and 3-amino-1,2,4 triazole. Endchain-modified poly(acrylates) or poly(styrenes) bearing thymine-endgroups were investigated in terms of melt behavior [91]. Here, melt viscosity was increased due to the hydrogen-bonding effects, but this was strongly dependent on the temperature. Thus, these materials may find application as rheological modifiers in industrial applications.

Binder et al. [92, 93] have reported on the formation of poly(etherketone) poly(isobutylene) networks formed by the respective endgroup-modified telechelics. The relevant interactions investigated relied on the 2,6-diamino-1,3,5-triazine/thymine and the much weaker cytosine/2,6-diamino-1,3,5-triazine-modified polymers (Fig. 18). In addition to the pure hydrogen-bonding interaction, phase-separation energies resulting from the strongly microphase separating PEK- and PIB polymers were expected. The association behavior was followed in solution via NMR-association experiments,

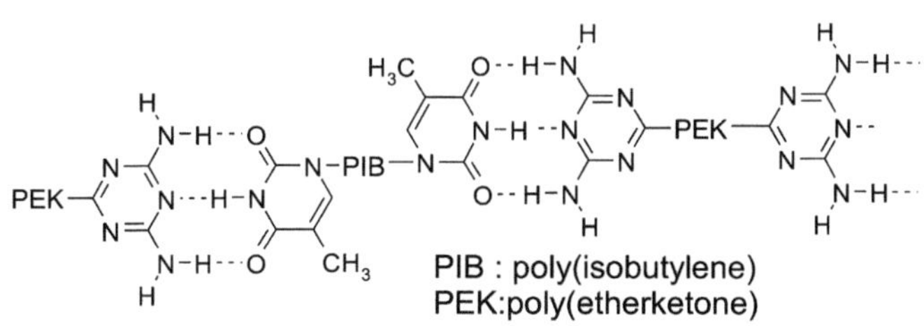

Fig. 18 Formation of pseudo-block copolymers consisting of poly(isobutylene)-poly(etherketone) telechelics held together by triple hydrogen bonds

revealing similar association constants of the hydrogen-bonding interactions of the polymers as compared to small molecular weight compounds. In the solid state, sheet-type structures are formed as studied by solid-state NMR spectroscopy, TEM- and thermal measurements. DSC methods clearly revealed the presence of two separate phases, whereas solid-state ^{13}C-MAS-NMR demonstrated the different chain mobility of the PEK- and PIB chains via relaxation measurements.

The relevance of thymine/2,6-diaminotriazine interactions has been exploited by a variety of authors to effect a reversible, yet stable association of catalysts, nanoparticles and other functional molecules onto polymeric molecules. Thus, Shen et al. [94, 95] reported on the formation of catalyst-supported structures for ATRP-polymerization via hydrogen-bonding systems (Fig. 19). The relevant Cu(I)-catalyst was affixed onto a poly(styrene) gel either via the thymine/2,6-diaminopyridine or the maleimide/2,6-diaminopyridine couple. The catalyst was able to mediate a living polymerization reaction of MMA in both cases, obviously acting in its dissociated form. The catalyst could be reused, retaining about half of its catalytic activity for further use. A strong solvent effect was observed, explainable by the dissociation of the catalyst from the support upon addition of strongly polar solvents.

The use of supramolecular interactions to bind a pharmaceutically active drug noncovalently to a polymer in order to achieve slow release was presented by Puskas et al. [96]. Here, a side-chain functionalized poly(styrene) bearing thymine moieties (Fig. 20) was prepared and complexed with phenol as a complexing agent. The release of the bound phenol was studied in aqueous buffer solution, revealing a slow desorption within 4.5 hours from the polymer. Thus, this system is adaptable for slow release of drugs from polymeric matrices.

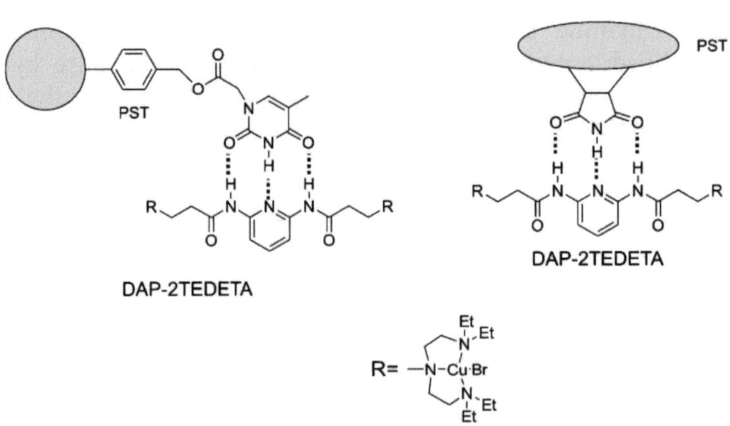

Fig. 19 Reversible attachment of a Cu(I)-catalyst to a solid support via triple hydrogen bonds, acting as a reversible catalyst for atom transfer radical polymerization (ATRP)

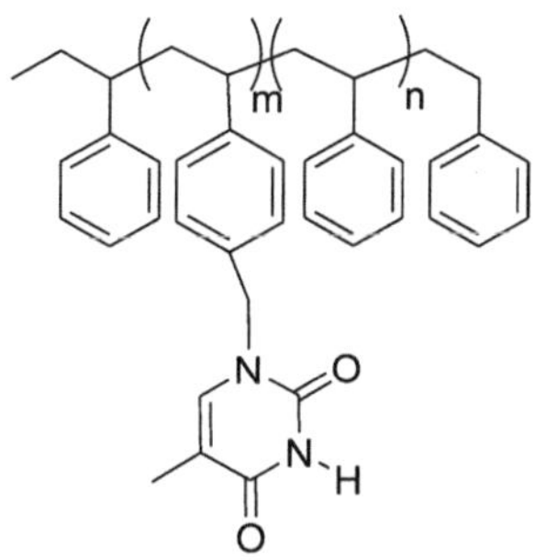

Fig. 20 Thymine-functionalized poly(styrene) for time-retarded drug release

A similar strategy for the binding of flavines was presented by Rotello et al. [97–99] (Fig. 21). Here, Merrifield-resins bearing side-chain functionalized poly(styrenes) with 2,6-diamino-1,3,5-triazines were prepared and subjected to the binding of flavines via triple hydrogen bonds. The concept has been modified for many different systems, using side-chain-modified poly(styrenes) [100], attaching systems such as nanoparticular structures (i.e.: POSS [101], Au-nanoparticles [102]) as well as redox-controllable systems such as ferrocenes [103]. The formation of aggregates such as polymeric microspheres [104] and polymersomes [105] have been reported with the same system (Fig. 21c). Thus, block copolymers consisting of norbornenes or poly(styrene) block copolymers bearing *N*-bisacyl-2,6-diamino-pyridine side-chains can be crosslinked with bivalent thymine derivatives. The crosslinking process can be followed by the incorporation of fluorescent dyes, revealing the structure formation in real time. Clearly, the multiplicity of medium-sized interactions ($K_{assn} \sim 200\ M^{-1}$) is the key-point for tuning the formation of these crosslinked structures. A related system has been extended to acrylates and poly(lactides), [106] where the recognition element is located within the central part of the polymeric chain. The complexation of the matching flavine residue has been followed by fluorescence spectroscopy, revealing an increase in the binding constant with increasing molecular weight of the flanking polymeric chains. This effect is explained by the differing of the average solvent concentration inside the volume enclosed by the polymers.

An excellent recent example of main chain liquid crystal formation was reported by Rowan et al. [107, 108] (Fig. 22). They have investigated liquid crystals held together via complementing nucleobases and a stiff bis(phenylethynylene)benzene core. The melt mixing of the corresponding matching building blocks generates materials with a much broader range of liquid crystallinity as compared to the individual components and additional fiber formation. A Hogsteen-type base pairing is proposed, leading to network formation and thus an increase in the overall aspect ratio and thus a more favorable formation of liquid crystallinity. A similar example

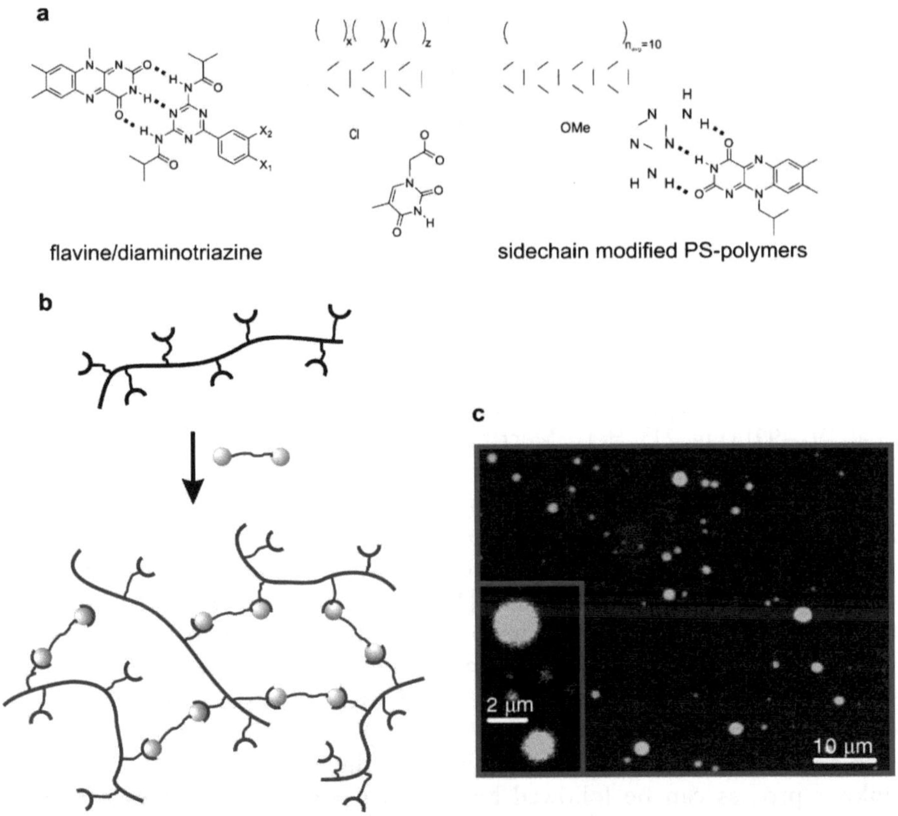

Fig. 21 Assembly of polymers via triple-hydrogen bonds. **a** Formula of the flavine/2,6-diamino-triazine and thymine/2,6-diaminotriazine interactions. **b** Formation of nanoparticles/polymer aggregates. **c** Polymersome formation by aggregation of poly(norbornenes) and poly(styrenes) bearing *N*-bisacyl-2,6-diamino-pyridine and thymine side-chains, respectively. Reprinted with permission from [104, 105]

generating liquid crystalline materials via hydrogen has been presented by the assembly of nucleobase bola amphiphilic structures [109] (Fig. 23). Here, bolaamphiphiles bearing thymine (T)- and adenine (A) nucleobases were assembled from solution. Thus, the complementing T-10-T and A-10-A molecules applied as a 1:1 mixture displayed the formation of nanometer-sized fibers instead of twisted- and helical ropes, as observed for T-10-T. The critical factor in determining the final structure was the molecular packing of the hydrophobic chains together with internucleobase interactions of the hydrogen bonds.

This internucleobase-crosslinking approach has been extended to the association of poly(tetrahydrofurane) segments (molecular weight < 2000 g mol^{-1}) by use of N^{6}-anisoyl-adenine or N^{4}-(4-tert-butylbenzoyl)cytosine as endgroups (Fig. 24) [110]. Despite the low association constant of these

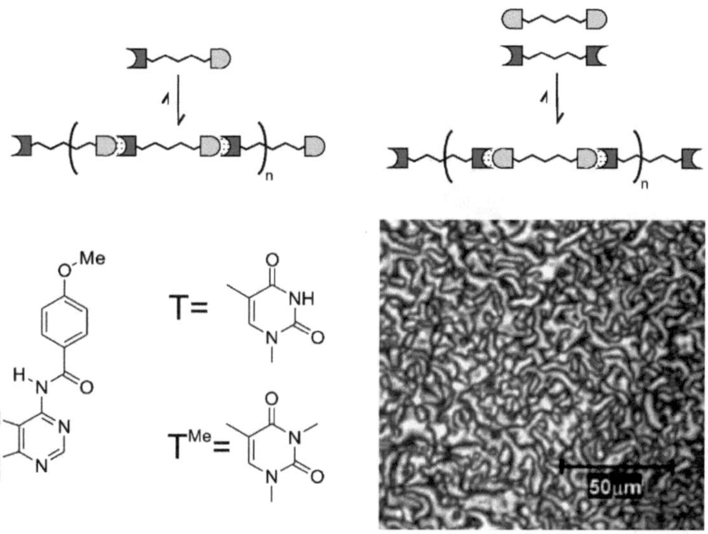

Fig. 22 Formation of liquid crystalline materials from main chain hydrogen-bonded polymers. Reprinted with permission from [107]

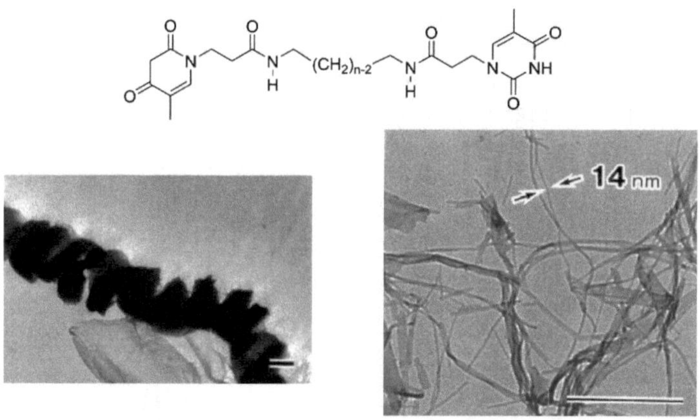

Fig. 23 Formation of fibers from bolaamphiphiles bearing thymine- or adenine end-groups. TEM micrographs of the fibers. Reprinted with permission from [109]

hydrogen-bonding moieties in solution ($K_{assn} < 5\ M^{-1}$) the soft PTHF forms materials with strongly different properties. Thus, a phase-separation process led to the formation of domains formed by crosslinked nucleobases, as well as soft-PTHF regions and was held responsible for the unusual material properties. Whereas PTHF is a waxy material with a melting point of around 20 °C, the resulting materials form flexible films as shown in Fig. 24. Thus, similar to the quadruple-hydrogen bonding systems discussed in the next section, even weak hydrogen bonding interactions can have a strong influence

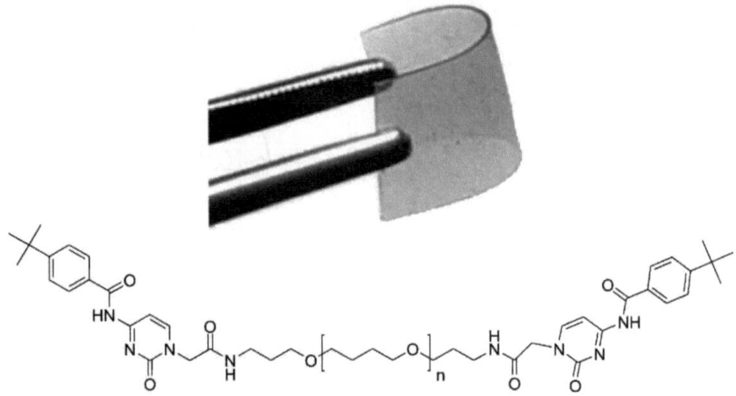

Fig. 24 Formation of flexible materials from poly(tetrahydrofurane) modified with N4-(4-tert-butylbenzoyl)cytosine endgroups. Reprinted with permission from [110]

on a materials properties, given that an efficient network formation can be achieved.

Finally, the complementarity of oligonucleotides provides a multiple hydrogen-bonding site with an extremely high selectivity, comparable to DNA. Using poly(styrene) endcapped with poly-thymidine (T_n) sequences, the microphase separation of the resulting films was investigated by Matsushita et al. [111] (Fig. 25). A synthetic procedure related to solid-phase synthesis of oligonucleotides was used to affix the final T_n-oligonucleotides to the poly(styrene) chain. Similar to the materials with single nucleobases,

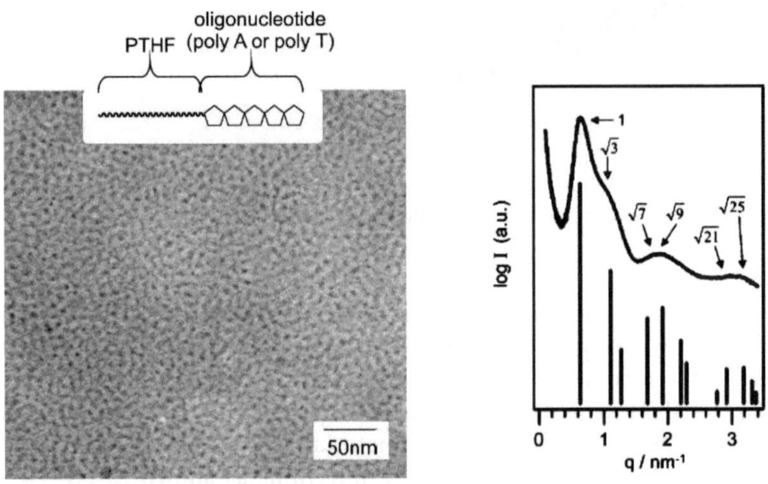

Fig. 25 Formation of microphase-separated films from poly(tetrahydrofuren) crosslinked with oligonucleotides. Reprinted with permission from [111]

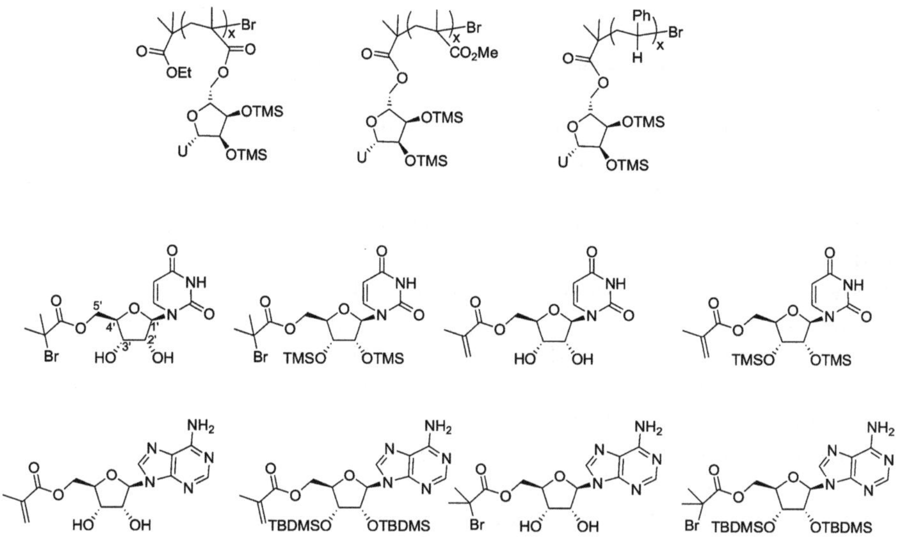

Fig. 26 Monomers and resulting homopolymers bearing nucleosides for template-recognition

microphase separation was observed, presumable due to network formation between the nucleobase strands, generating individual PS- and nucleoside domains. SAXS-investigations revealed microphase-separated structures with a high order and domain spacings of about 11.4 nm.

A very interesting approach to use of the recognition abilities of nucleosides linked to polymers was reported by Haddleton et al. [112, 113] (Fig. 26). Homopolymers bearing the poly(A) or poly(U) nucleosides were prepared by radical polymerization. These homopolymers were then used as templates to direct the polymerization of the preferentially matching monomeric units and thus guide the formation of homopolymers in relation to statistical mixtures of either monomer within the copolymer via free radical polymerization. Thus, it was shown that poly(U) could act as a template for the polymerization of an adenine-containing monomer. However, complexation phenomena strongly inhibited a detailed analysis of all templating effects, presumable due to Hogsteen-base pairing.

The self-recognition abilities of 2,6-diaminoacyl-pyridine (DAP) with thymine derivatives was exploited recently using an imprinting approach [114]. Here, a diblock copolymer consisting of poly(tert-butyl methacrylate)-*b*-(2-hydroxyethyl methacrylate) was prepared via ATRP-methods and subsequently derivatized with DAP-units. After the formation of block copolymeric micelles in a selective solvent with the matching thymine, the ligand was removed. The resulting particles were then able to interact selectively with the matching thymine-, but not with the nonmatching *N*-alkylated thymine derivatives.

3.4 Polymers Connected with Quadruple Hydrogen Bonds

Quadrupolar hydrogen bonds have been designed to enhance the effects studied and known from mono-, double and triple hydrogen bonds. A thorough overview over most known quadrupolar hydrogen-bonding systems is given by Bhattacharya et al. [28]. The association of small building blocks leading to materials with polymeric properties in solution is restricted by the strength of the binding constant as calculated by Meijer et al. [11] A binding constant at 10 000 M^{-1} leads to an aggregation number of about 100 molecules at a concentration of the building blocks at 1 M. In order to reach a sufficient virtual chain length, an association constant of more than 10 000 M^{-1} is advisable. Thus, the 2-ureido-4[1H] pyrimidinone unit [21] (and related units such as the ureidotriazine unit [115]), (Fig. 27) were developed featuring a binding constant of 6×10^7 M^{-1}, leading to a very strong association of the building blocks. Several reviews have dealt with the topic of quadrupolar hydrogen bonding [9, 14, 116, 117].

The first publication describing these strong effects was in 1999 by Lange et al. [118] (Fig. 28). They used the dimerizing ability of the ureidopyrimidine units to generate reversible polymer networks composed of PEO- and PPO-telechelics. The introduction of the ureidopyrimidine moiety was accomplished easily via terminal isocyanates and subsequent reaction with methylisocytosine—a method which also allows the industrial scale-up via easily available hydroxy-telechelic polymers and bivalent isocyanates, such as TDI, MDI and isophorone-diisocyanate [119]. The well-defined dimerization of the ureidopyrimidine moiety allows the formation of a network, not requiring additional stabilization such as crystallization or phase separation of the polymeric components. The resulting material displays a well-defined viscoelastic transition. Addition of water (up to 11% w/w) caused a significant, but not entire drop of the viscosity in solution, whereas the addition of monomeric units bearing the ureidopyrimidine moiety entirely dropped the favorable properties of the polymer. Thus, the quadruple hydrogen bonds still

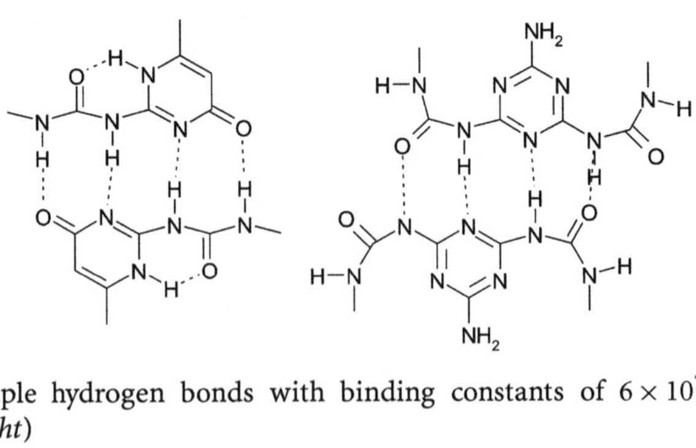

Fig. 27 Quadruple hydrogen bonds with binding constants of 6×10^7 M^{-1} (*left*) and 2×10^4 M^{-1} (*right*)

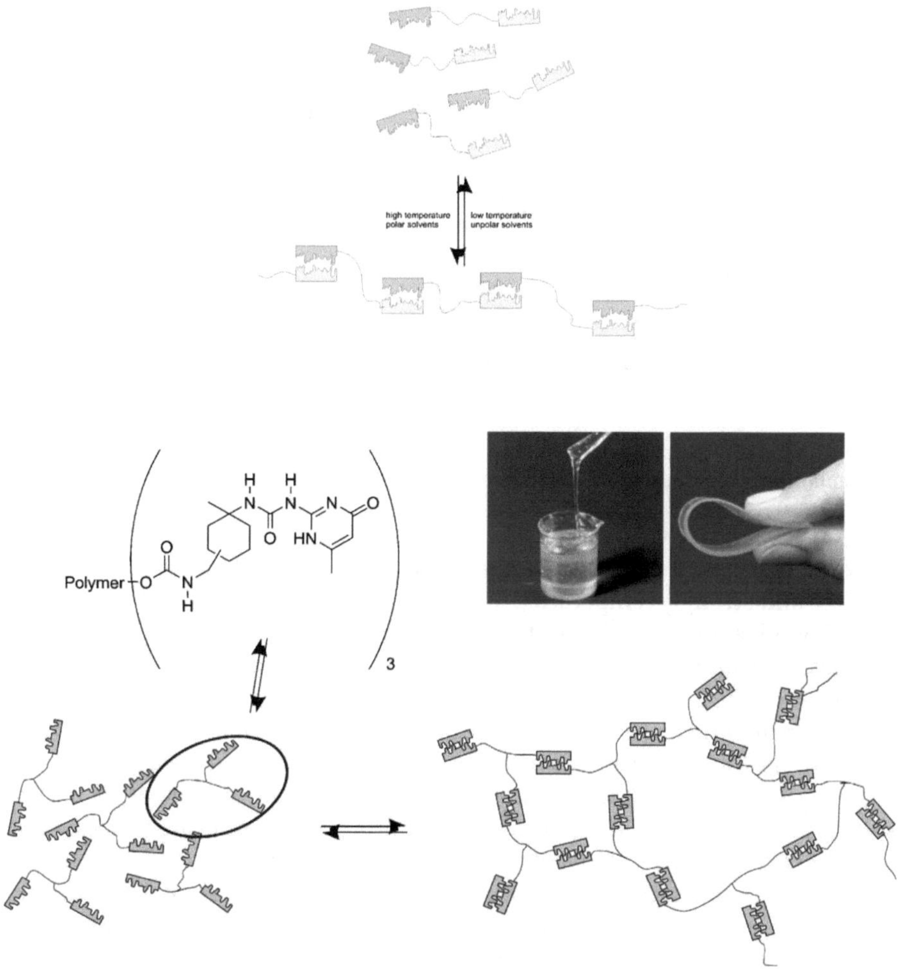

Fig. 28 Formation of supramolecular polymers and networks via self complementing quadrupolar hydrogen bonds. Reprinted with permission from [118]

display stability even in the presence of this large amount of water, but are readily cleaved by interaction with monomeric units, thus leading to a strong depolymerization of the material.

Meijer et al. [120] have also related the virtual degree of polymerization of the associating building blocks to the interaction strength of the respective endgroups (Fig. 29). A higher degree of association can either be achieved by a stronger association constant, or via a higher concentration of the building blocks. Since the concentration of the building blocks is limited by solubility, the association constant is supposed to be at least 10^5 M^{-1} in order to reach a degree of association of ~ 100 at a concentration of the building blocks of ~ 0.05 M. This explains the high potency of Meijers quadruple hydrogen-

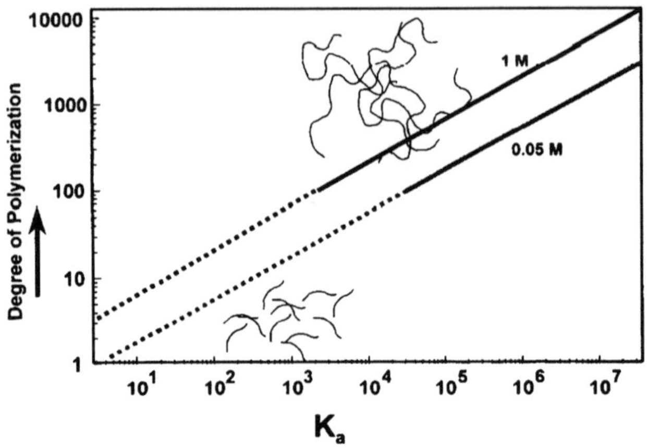

Fig. 29 "Virtual" degree of polymerization versus association strength (K_a) in supramolecular polymers. Reprinted with permission from [120]

bonding systems. Especially for use in coatings and hot melts, where a reversible and strongly temperature-dependent rheology is required, the system is highly advantageous. In particular, the unidirectional hydrogen-bonding systems, although self-complementary, prevent multidirectional association and gelation [121].

Within the quadrupolar hydrogen-bonding systems, three aspects are significant: (1) the action of the association in diluted liquid solution, (2) the ordering in the solid state, most of all the mechanical properties of the associates, and (3) methods to investigate and prove the ordering process. The latter is strongly related to electric- and optical properties.

In solution an important aspect concerns the equilibrium between aggregation into chains or medium to small-sized rings. Since the formation process of supramolecular polymers is comparable to the polycondensation reactions, a similar approach in the sense of Jacobson and Stockmayer [122] has been applied to this question. Thus, different chain length separating dimeric ureidopyrimidines were studied in solution, focussing on their ring/chain equilibria [123]. A strong influence of flanking methyl substituents together with the chain length [124] was found to dominate the formation of either rings or chains. A critical concentration was detected, above which the amount of cyclic structures remains constant. Additionally, the system can be used to assemble only homochiral dimeric structures into rings [125]. Since similar to poly(condensation) reactions the ring formation strongly decreases the molecular weight of the formed systems, this aspect is important in the design of new supramolecular structures, either favoring the ring- or the chain formation.

Another effect studied intensely in solution concerns the reversible switching of association-equilibria by photochemical effects. Blocked o-nitrobenzyl derivatives of ureidopyrimidones have been used for this purpose [126], enabling the cleavage of the photolabile o-nitrobenzyl moiety from the ureidopyrimidone unit, thus allowing the generation of a monovalent species able to inhibit the aggregation process, and thus reduce the viscosity of the mixture by pure irradiation. Another more sophisticated approach placed the photoswitchable structure directly in the main chain of the spacer between flanking ureidopyrimidine units (Fig. 30) [127]. Thus, the photochromic dithienylethylene was chosen as the photochemically active moiety, changing between the closed and open structure via a [3,3]-Cope cyclization. Irradiation at 366 nm led to the closed form, whereas irradiation at > 540 nm led to the initial open form. Since only the closed form is able to aggregate, there is a strong increase in aggregate formation upon irradiation at 366 nm, which is reversible upon irradiation at > 540 nm. Thus, a switchable system has been generated, able to define aggregate size and thus viscosity upon simple irradiation.

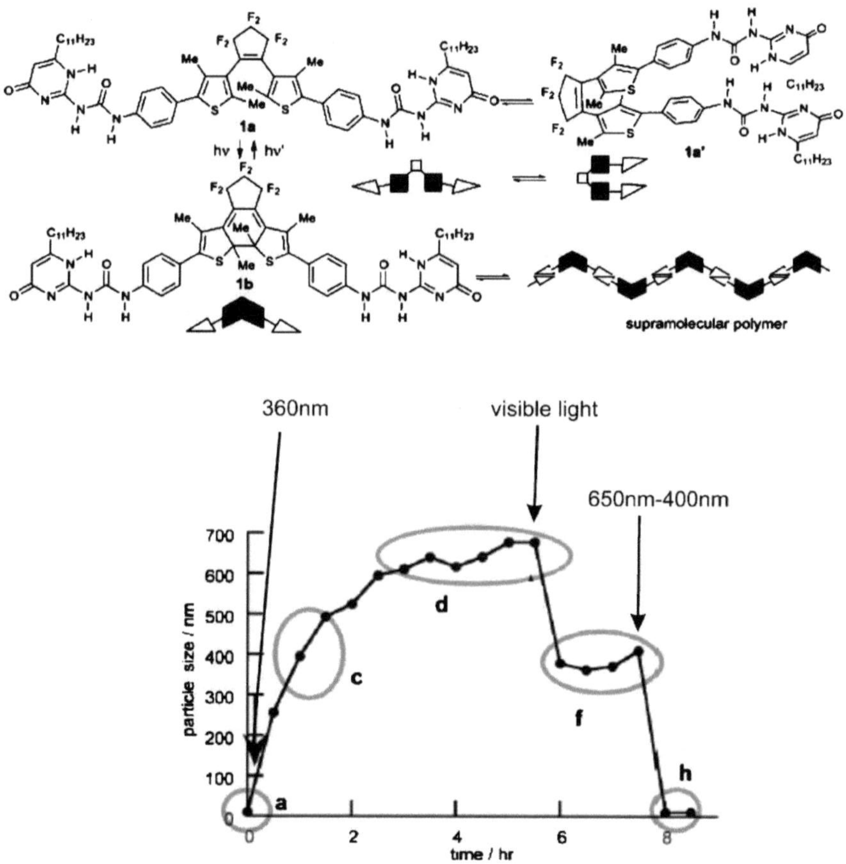

Fig. 30 Phototriggered aggregation of dithienylethylenes with flanking ureidopyrimidone-units. Reprinted with permission from [127]

Significant effects in solution were observed upon assembling pi-conjugated polymer segments with fluorene oligomers and perylene-oligomers (Fig. 31a) [128]. The association of the building units is achieved from solution, achieving energy transfer due to the stacking of the units, transferring energy from the (blue-absorbing) oligofluorenes to the (green-emissive) phenylene-vinylenes. The ordering permits a more efficient energy transfer process due to the neighboring of the functional groups. Similar systems work with assembling the oligophenylenes into discs (Fig. 31b) [129] and lad-

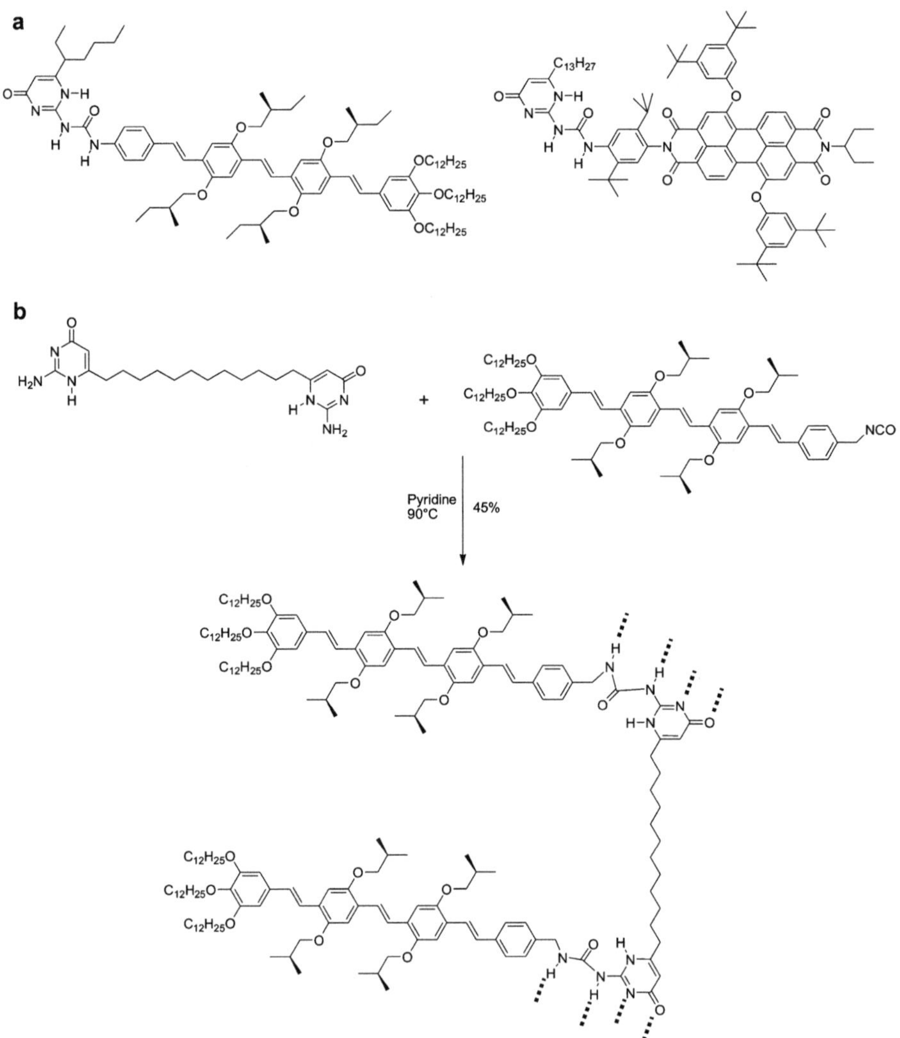

Fig. 31 Self association of **a** oligo(p-phenylenevinylenes) and perylenes as well as **b** olig(p-phenylenevinylenes) via quadruple hydrogen bonds

ders [130] that show a strong exciton coupling in solution. Furthermore, the application of these solutions together with fullerenes permitted the formation of processable thin films, able to control the film morphology via spin coating. Neutron-scattering experiments on the resulting films with thicknesses of about 100 nm indicated a rod-like structure similar to those in solution, able to blend the fullerene derivative close to the oligo(phenylene) structure. The resulting photovoltaic devices showed a photo-to-current transformation of about 12%, presumable due to the phase-separated structure of the film.

Studies in the solid state and molten state reveal dramatic changes of the resulting materials with respect to the unmodified polymers. The system has been successfully applied to a variety of oligomeric and polymeric structures. Sijbesma et al. [131] have investigated the melt viscosities of various telechelics [i.e.: poly(ethylene-butylene), poly(ethyleneoxide-propyleneoxide)], polyesters with molecular weights ranging from ~ 2000 to $3500\,\mathrm{g\,mol^{-1}}$. They found a complex thermorheological behavior (Fig. 32), which can be explained by the reversibility of the hydrogen bonds. Clusters are formed, consisting of hydrogen-bonded structures, and these were explained to act as physical crosslinks, similar to those in thermoplastic elastomers. Melt viscosity strongly decreases with temperature (i.e.: the melt viscosity decreases by a factor of 4 for a temperature increase of 10 °C). Similar strong behavior of the melt viscosities was demonstrated by Long et al. and others with ureidopyrimidine-bound acrylates [132] (Fig. 33), oligo-siloxanes [133], poly(esters) [134]. Similar, very strong effects have been observed with side-chain-modified ureidopyrimidones affixed onto poly(hexene) (Fig. 34) [135]. Thus, at low concentrations of the ureidopyrim-

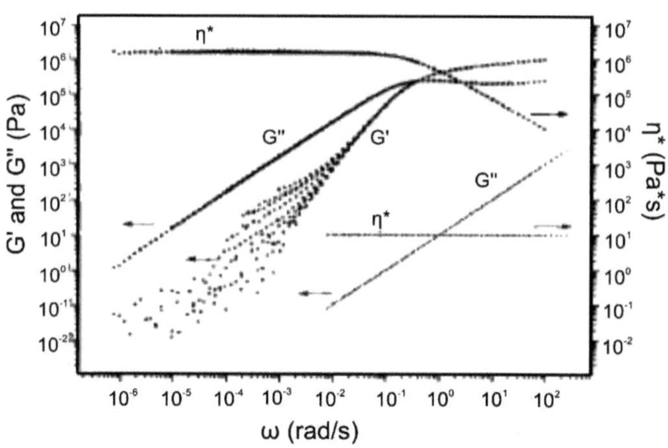

Fig. 32 Master curve of the dynamic melt viscosity (η^*), storage modulus (G′) and loss modulus (G″) of a poly(ethylene/butylene) copolymer ($M_n = 3500\,\mathrm{g\,mol^{-1}}$). Reprinted with permission from [131]

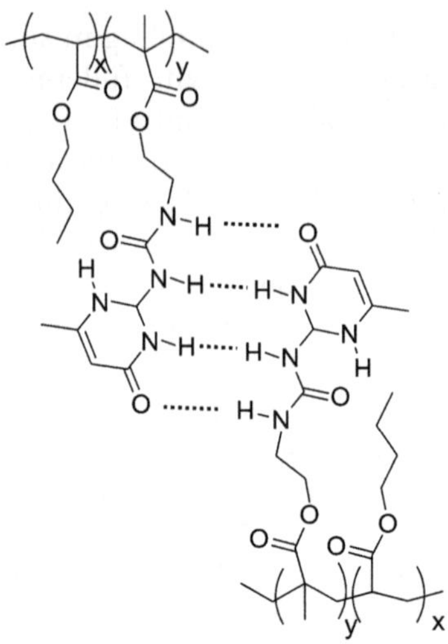

Fig. 33 Association of poly(acrylates) via pendant ureidopyrimidone groups

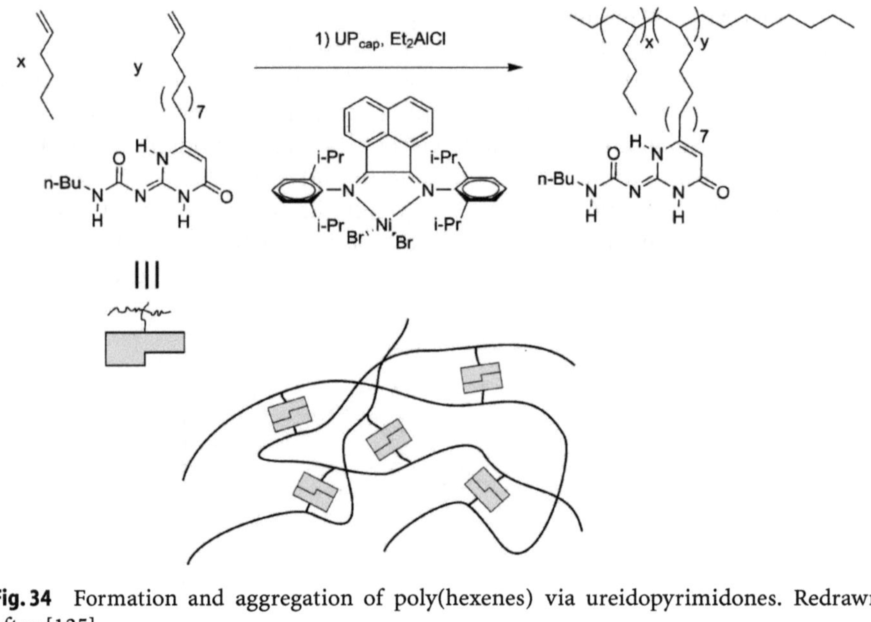

Fig. 34 Formation and aggregation of poly(hexenes) via ureidopyrimidones. Redrawn after [135]

idine unit (below 3 mol %) the resulting copolymer is comparable to the corresponding homopolymer without the hydrogen-bonding systems. At higher concentrations, the reversible crosslinks are very effective, transforming the polymer to an amorphous network with elastomeric properties: thus much higher stress values are observed at rupture, presumably due to the action of the hydrogen-bonding systems preventing the chains from moving apart.

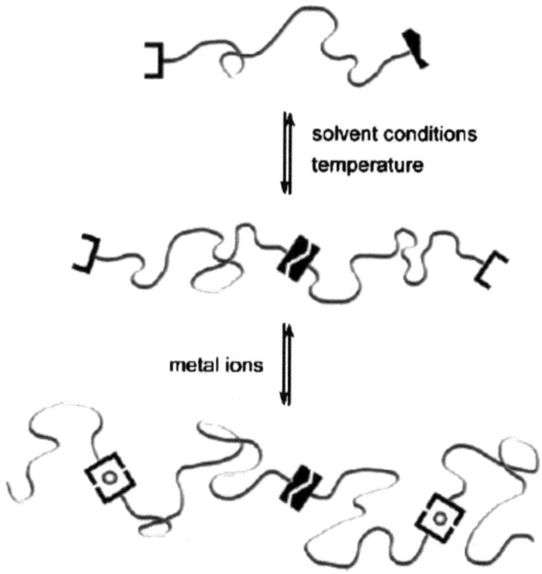

Fig. 35 Combined association via quadrupolar hydrogen bonds and metal complexes. Reprinted with permission from [136]

A combination of two different supramolecular forces, namely metal–metal bonds and the ureidopyrimidones were investigated by Schubert et al. [136, 137] (Fig. 35). A poly(ε-caprolactame) was prepared—with a terpyridine moiety at one end and an ureidopyrimidone unit at the other end—via tin-octanoate catalyzed polymerization. Together with iron(II) ions, double supramolecular polymers formed from chloroform solution. Again, solid-like behavior was observed, similar to results for the purely hydrogen-bonded polymers.

The ability to organize supramolecular polymers held together by quadruple bonds into fiber-like structures was exploited by Meijer et al. to construct bioactive scaffolds for tissue engineering (Fig. 36) [138]. Mixtures of ureidopyrimidone polymers bearing either biocompatible poly(ε-caprolactam) as well as peptidic recognition sequences (GRGDS) for mediating cell attachment, were processed from solution and spun into fibers. The fibers then consisted of the biocompatible polymer and the cellular-recognition sequence, able to harvest cells after incubating the readily spun material into its final shape. After cellular seeding and growth, the material dissolves slowly into the aqueous phase due to either proteolytic cleavage or the breakage of the hydrogen bonds. Thus, a new material acting as a cellular scaffold and with excellent resorption qualities has been generated.

The generation of supramolecular block copolymers certainly is among the most important topics for a polymer chemist. Since the ureidopyrimidone moiety is a self-complementing hydrogen bond, this issue is difficult to ad-

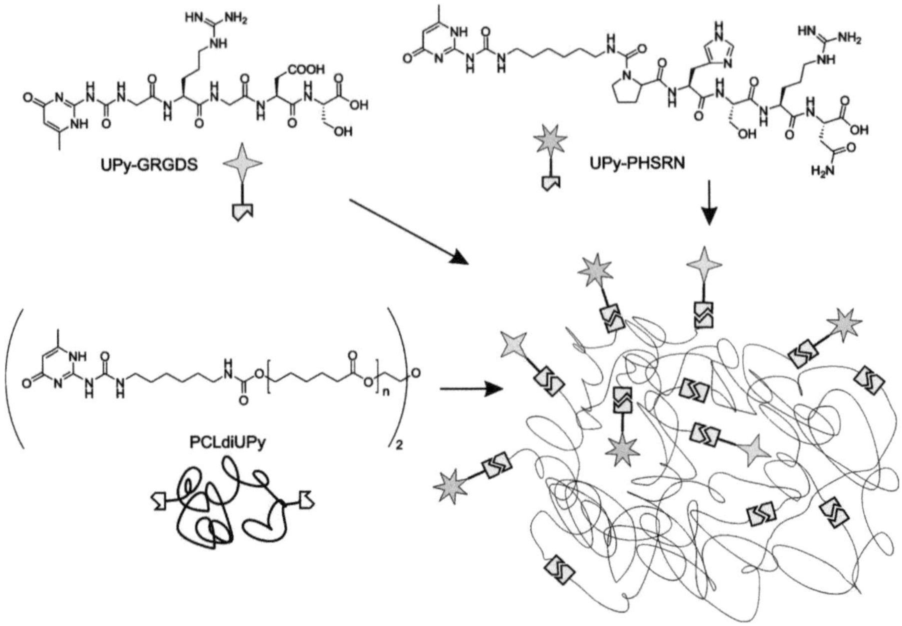

Fig. 36 Formation of bioactive, fiber-scaffolds for tissue engineering. Mixtures of telechelic ureidopyrimidone polymers [consisting either of poly(e-caprolactam) or peptidic GRGDS sequences] are processes into fibers

dress. In principle, there are two modes of approach: start from a block-di- or triblock copolymer and make it assemble into the right A–B mode as one would assume due to microphase separation, or use two different quadrupolar hydrogen-bonding systems, that can effect a clean A–B-type interaction. Meijer et al. have chosen the latter approach by generating a different set of hydrogen-bonding interaction (Fig. 37) [139]. Thus, the two different hydrogen-bonding systems match each other only in one case, putting the poly(tetrahydrofurane) and alkyl-segment into a defined order. The new quadrupolar interaction favors the ureidopyrimidone/2,7-diamido-1,8-naphthyridines interaction over the respective self-complementing interaction. In a similar sense, the pure ureidopyrimidone system can be used to generate block copolymers via the self-complementing ureidopyrimidone bonds. [140]. Telechelic block copolymers [poly(styrene)-block-poly(*cis*-isoprene)] have been produced with affixed ureidopyrimidines as endgroups.

Polymerization-induced phase separation (PIPS) within hydrogen-bonded polymeric fragments can lead to materials with combined properties and defined morphology. Keizer et al. [141] (Fig. 38) described the formation of "droplets" within an acrylate matrix using a miscible supramolecular polymer bearing associative endgroups. The miscibility is reduced during the polymerization reaction of the acrylate, thus leading to finely disperged

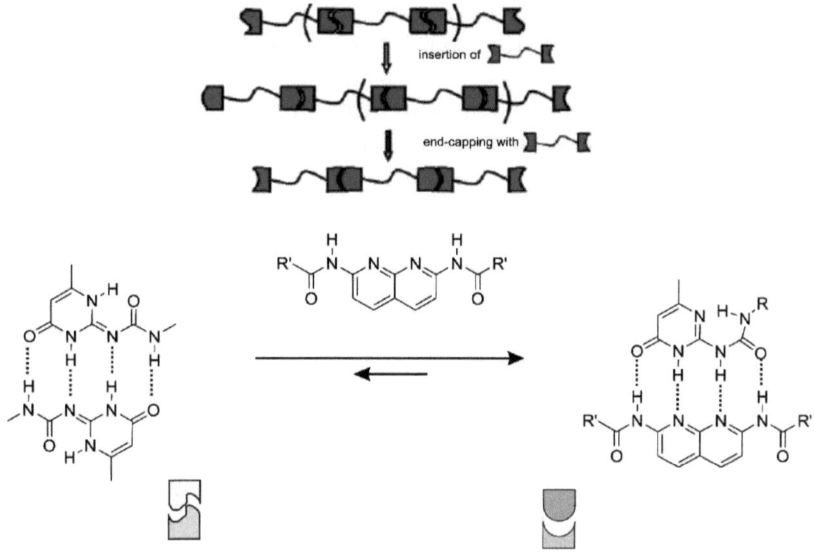

Fig. 37 Generation of block-copolymers via changing the association pattern by addition of 2,7-diamido-1,8-naphtyridines. Reprinted with permission from [139]

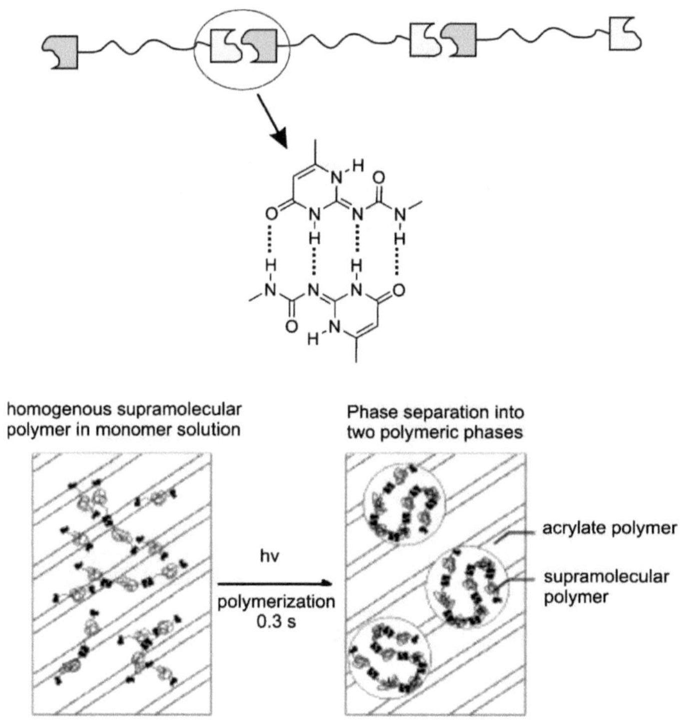

Fig. 38 Polymerization-induced phase separation of a supramolecular polymer. Reprinted with permission from [141]

droplets of the supramolecular polymer within the bulk-polymer matrix. The size of the droplets can be varied starting with less than 100 nm up to several microns in diameter.

Defined surface roughness can be induced by hydrogen-bonding systems bearing trialkoxy-siloxane units (Fig. 39) [142]. The hydrogen-bonded asso-

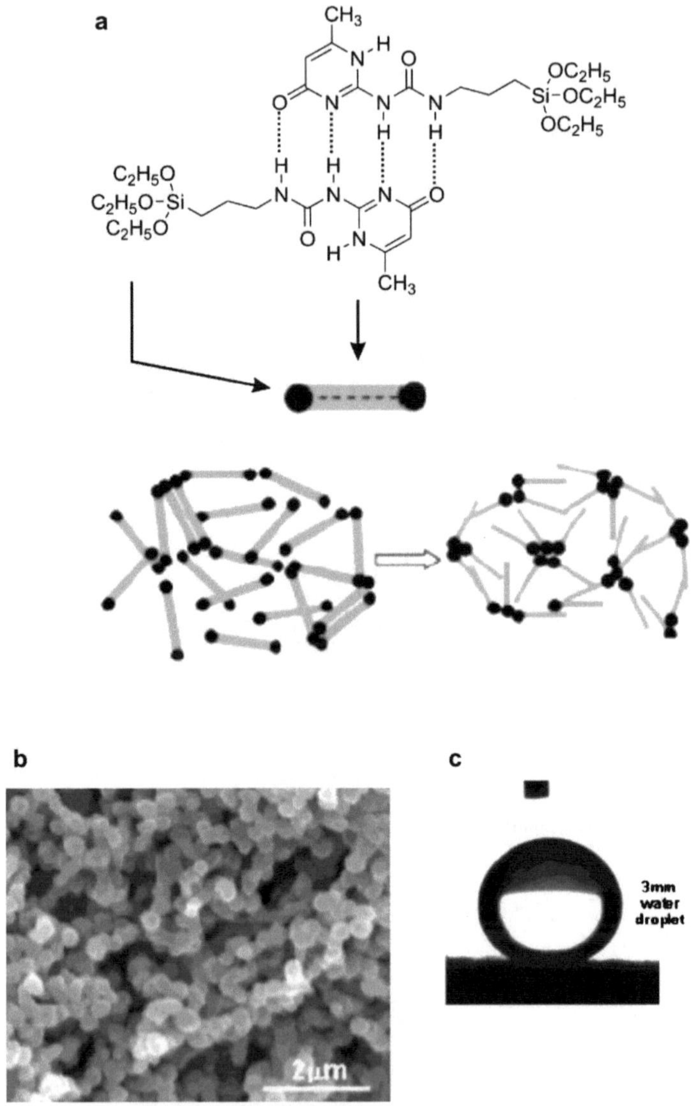

Fig. 39 Generation of defined surface roughness during a sol/gel process. **a** Concept to introduce quadrupolar hydrogen bonds into a silica gel. **b** Surface roughness via SEM. **c** Superhydrophobicity of the resulting surface. Reprinted with permission from [142]

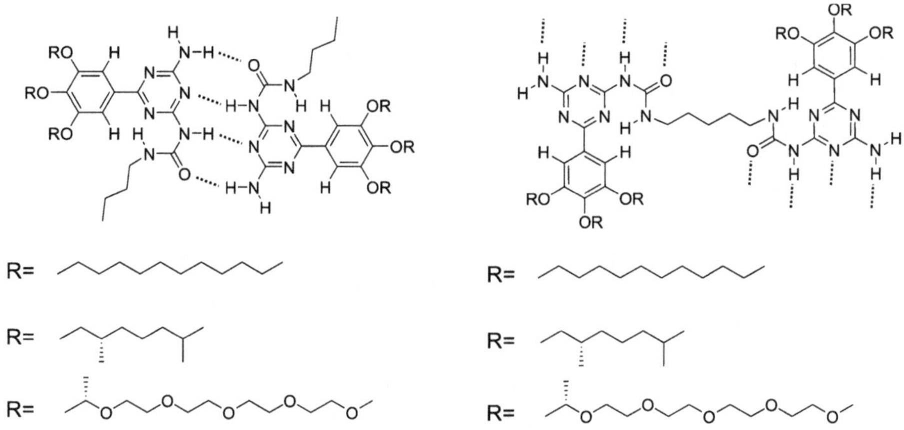

Fig. 40 Supramolecular polymers formed by ureidotriazines

ciate is subjected to a sol–gel-mediated crosslinking process of the siloxane units, leading to a three-dimensional network. Phase-separation processes lead to the formation of defined granules and holes on the surface. The material displays a defined surface roughness and thus superhydrophobic contact angles in contact with water, mimicking the self-cleaning properties of the lotus leaf [143, 144]. Thus, the noncovalent preorganization on the basis of the hydrogen bonds leads to phase effects not reachable with conventional polymer systems and presents a biomimetic approach to a highly sophisticated, bioinspired material.

Besides the aforementioned hydrogen bonds, another related system has been developed by Meijer et al. (Fig. 40) [145]. Related ureidotriazines can be assembled in a similar mode as the previously mentioned ureidopyrimidones. This type of hydrogen bond is self complementary, with a binding constant of about 2×10^4 M^{-1}. Association into columns is observed, showing effects similar to those of the ureidopyrimidone quadrupolar hydrogen bonds. Well-defined helices can be formed from within these associates [146]. Since these endgroups can be easily fabricated by reacting an isocyanate with 1,3,5-triamino-2,4,6-triazine, these bonds are of significant industrial importance. Similar drastic effects on material properties were observed with poly(tetrahydrofurane) supramolecular compounds held together via this bond [147].

3.5 Polymers Connected with Multiple Hydrogen Bonds

Multiple hydrogen bonds are an extension of the concept to introduce more connecting sites within a hydrogen-bonding system. Several reasons have driven the quest for these higher-order systems for the construction of new polymeric architectures and structures: (a) stability enhancement by adding more hydrogen bonds into the binding systems, thus enhancing the crosslinking effects in the final solids or the primary solution; (b) the quest for network formation, instead of only generating linear systems via additional crosslinking due to the additional hydrogen bonds, enabling more interactions to neighboring chains; and (c) the need to avoid ureidopyrimidones due to their self-complementary structure. The latter is the most important, since the ureidopyrimidones, despite their enormous effects and capacity, cannot be used to combine two different materials in an alternating manner and thus achieve sequence control.

One of the most important hydrogen-bonding systems used as an alternative system to the ureidopyrimidones, able to order in a heteroassembling manner, is comprised of a hexapolar hydrogen-bonding moiety [DAD–DAD (D = donor; A = acceptor)] named after Hamilton et al. [32, 148, 149] with either barbituric acid or *N*-alkyl-cyanurate. This multivariant receptor has shown a pronounced type of heteroassembly, directing the molecules in a defined heterocomplementary (A-B-type) fashion. The binding constant is about $10^5\ M^{-1}$, putting it close to those of the (homo-associating) ureidopyrimidones. The system was initially chosen by Lehn et al. to mediate the assembly of both bivalent and trivalent building blocks into fibers, gels and networks [150]. Thus, the building blocks shown in Fig. 41 can assemble into linear structures, which leads to fibers upon higher association. Several effects leading to the formation of fiber-type structures were studied. (a) An

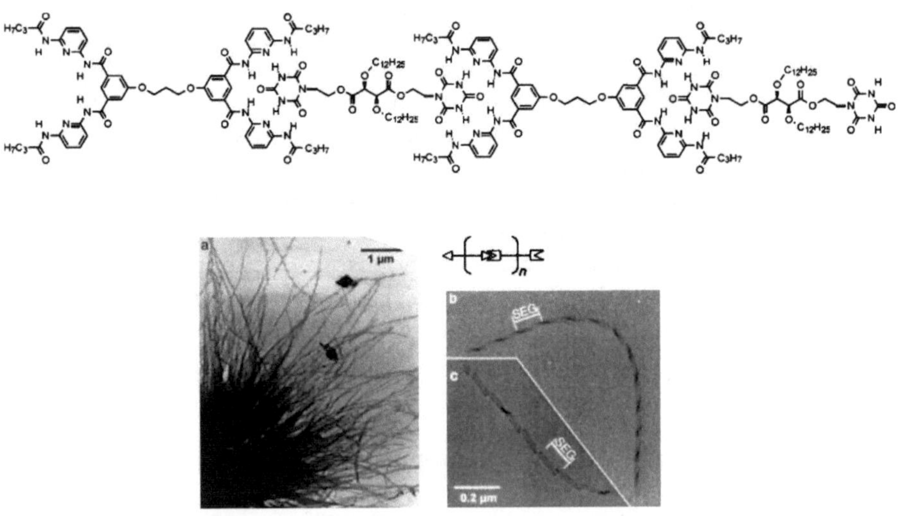

Fig. 41 Formation of helices and fibers from low molecular weight building blocks. The association takes place via the Hamilton receptor/cyanurate interaction. Reprinted with permission from [150]

increase of the viscosity due to the strongly enhanced molecular weight. (b) Proof of the reversible association by temperature-dependent measurements of the N – H chemical shift in ^{1}H-NMR spectroscopy. (c) Aggregate formation was studied by ^{1}H-NMR relaxation measurements monitoring the change in T_1-relaxation time, which led to an increase of T_1 of the aggregates with respect to the monomolecular species. (d) Most importantly, the concentration dependence of the association in different solvents with and without a monovalent "chain-stopper" and imbalanced stoichiometric ratios of the two components were studied. It was demonstrated, that an already small imbalance in stoichiometry between the two assembling components led to a strong drop of the virtual molecular weight of the associates, in accordance with simulations related to polycondensation reactions. Gel and fiber formation was proven by electron microscopy, revealing bundles of fibers. As demonstrated later by neutron-scattering experiments in nonpolar solvents such as decane or toluene [151], the formation of microgels, consisting of interlocked bundles of helices was proven. Thus, the interlocking of several helical strands leads to partial crosslinking under certain conditions, presumably via additional aromatic stacking effects. The formation of helical strands can be achieved (Fig. 42) via induced recognition by specific hosts, being able to induce conformational restraints within alternating chains of hydrogen-bonding units. The formation of specific rotamers leads to a specific folding of the chain, and thus subsequently to the formation of helical strands, which again can be visualized as fibers, consisting of helical aggregates [152, 153].

The aggregation using the aforementioned hydrogen bonds has been extended to more sophisticated building blocks. Zhuang et al. [154] studied the aggregation of a pincer-like molecule (such as a perylene-bisimide bearing two "Hamilton"-receptors on either side) and a [60]fullerene derivative with a matching barbituric acid derivative (Fig. 43). The aggregation was proven via ^{1}H-NMR spectroscopy and—most importantly—via the fluorescence quenching of the perylene by the fullerene derivative. Because of the supramolecular organization, the molecular proximity of the substrates is close as revealed by the intermolecular charge transfer. It was possible to indicate photo-induced electron transfer between the perylene and the [60]fullerene, presenting the possibility to build new photovoltaic devices with nanoscale structures.

The concept of association in solution via the sextuple hydrogen bonds, combined with a (partially) reversible covalent bond (mostly imines) [155] has led to the concept of dynamic combinatorial libraries, or so-called double dynamers (Fig. 44) [156]. The underlying component exchange due to the dynamic bonds allows the generation of constitutional diversity of both the molecular and the supramolecular level. Thus, selection processes may be both driven by the hydrogen-bonding interaction, as well as the partially reversible covalent bonding system. New materials with strongly modulative properties can be envisioned from this concept.

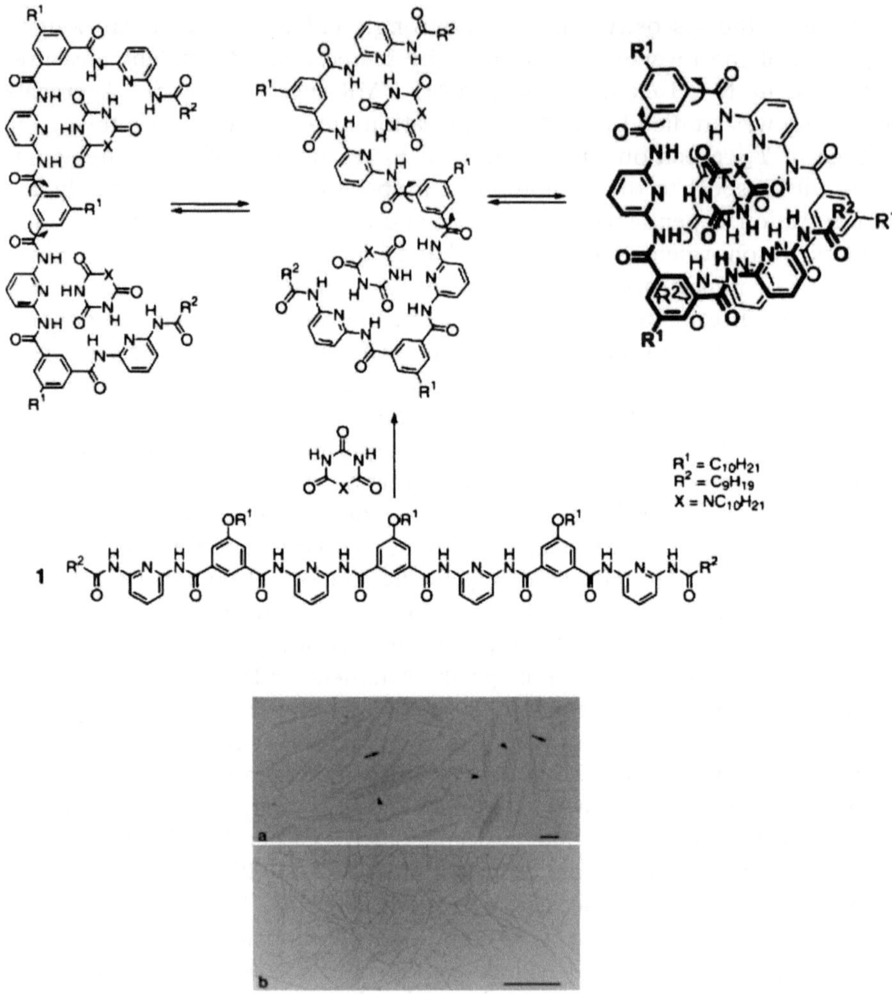

Fig. 42 Formation of helical strands by selection of specific rotamers during the assembly process. Reprinted with permission from [152]

The combination of strongly differing polymeric or oligomeric building blocks has been achieved by use of the Hamilton-receptor/barbituric acid systems, generating pseudo-block copolymers linked together by purely non-covalent bonds (Fig. 45) [157]. Thus, two strongly microphase-separating polymers [poly(etherketone) [158] and poly(isobutylene) [159]] with the respective hydrogen-bonding systems were combined in a solution blending approach. The resulting materials differed strongly in their appearance from their individual components, revealing two separate glass-transition temperatures. Microphase separation was proven by TEM and SAXS investigations, as well as via solid-state NMR experiments [160]. Temperature-dependent

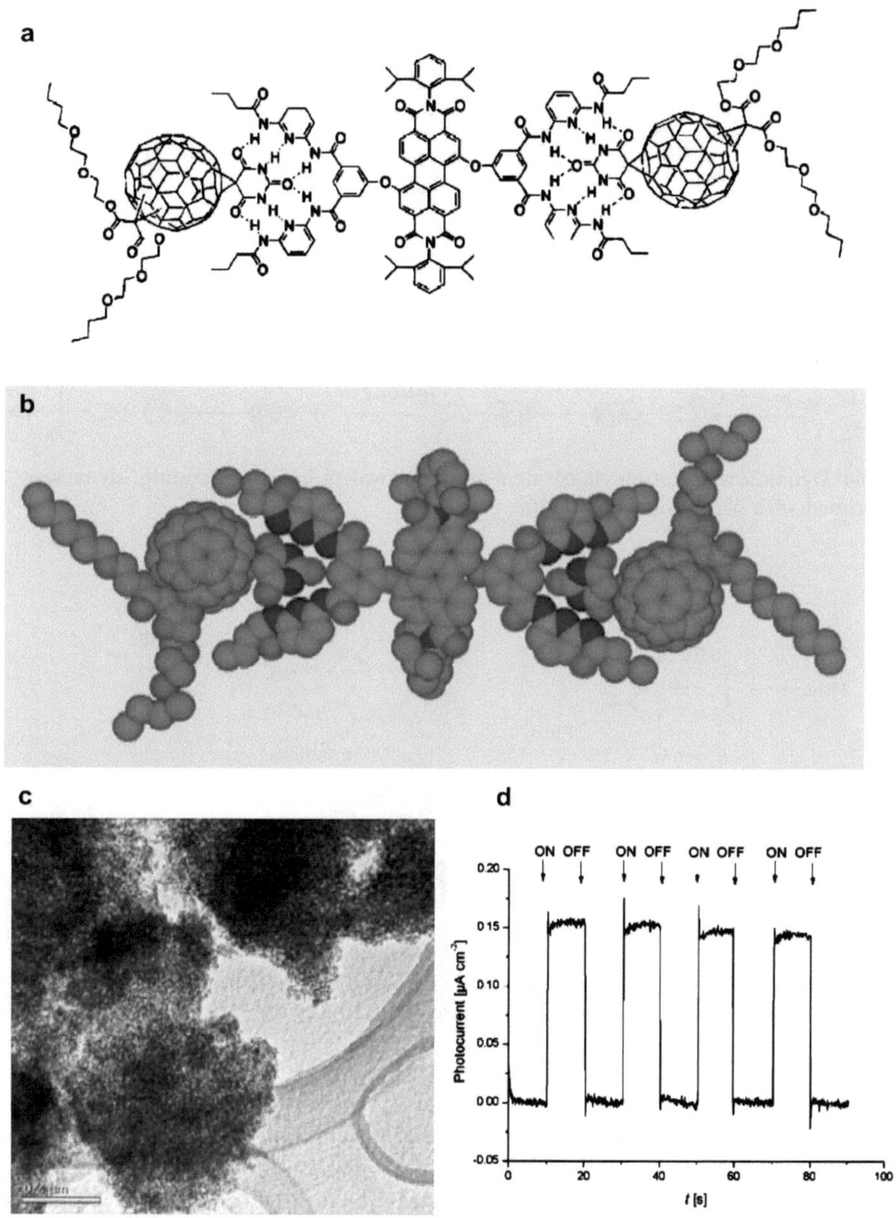

Fig. 43 **a** Perylene/fullerene aggregates. **b** Steric 3D-representation of the assembly. **c** TEM micrograph of the formed aggregate. **d** Photocurrent generation upon irradiation. Reprinted with permission from [154]

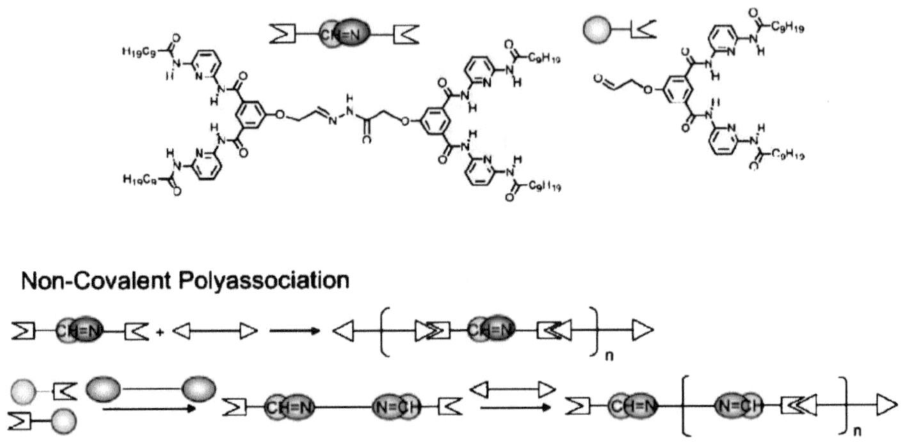

Fig. 44 Dynamic association via covalent and noncovalent bonds generating "dynamers." Reprinted with permission from [156]

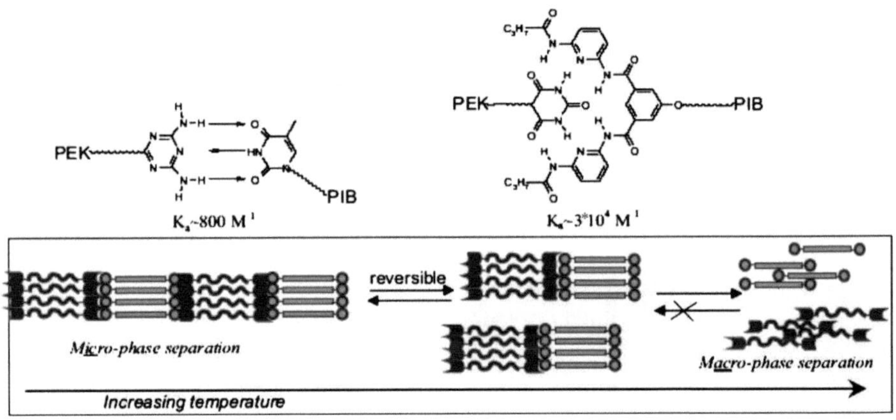

Fig. 45 Pseudo-block copolymers associated via multiple hydrogen bonds and their temperature behavior in correlation to microphase separation of the PIB and PEK-telechelics, respectively

SAXS measurements revealed a strong dependence of the microphase separation in relation to the strength of the hydrogen-bonding interaction. Thus, the sextuple-hydrogen-bonding system was able to glue the polymeric phases together up to temperatures 80 °C above the glass-transition temperature of the higher melting component. The concept has been extended to gels, where trivalent structures mediate the formation of thermoreversible gels [161]. A similar approach with higher-order bonding systems has been described by Ryu et al. generating diblock copolymers [162]. The two immiscible polymers poly(styrene) and poly(ethyleneglycol) were coupled to the polyamidic hydrogen-bonding structure with an association constant of $K_{assn} > 10^9\ M^{-1}$

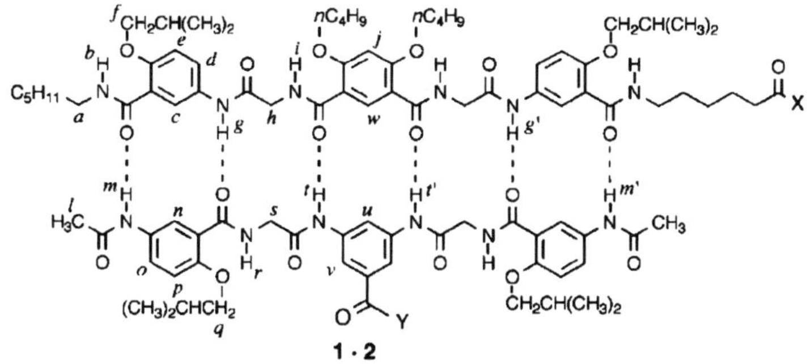

1
2

1 · 2

X = -OCH$_3$
-O-PS (M_w = 1.7×10^3)
-O-PS (M_w = 3.2×10^3)
-O-PS (M_w = 20×10^3)

Y = -OC(CH$_3$)$_3$
-O-PEG (M_w = 350×10^3)
-O-PEG (M_w = 2.0×10^3)
-O-PEG (M_w = 5.0×10^3)

Fig. 46 Combination of two immiscible polymers (PS and PEG) via multiple hydrogen bonds

and merged in a solution blending approach (Fig. 46). Again two separate glass-transition temperatures were observed, together with the microphase separation proven by AFM investigations of thin films.

It has been recognized that multiple hydrogen-bonding systems offer the possibility to form networks, which can form strands, sheets and rosettes (for a general overview see: [163]). The common system is derived from 2,4,6-triaminotriazine (melamine)/barbituric acid, and leads to an extremely high level of order, tunable by the patterns of substituents, hydrophobicity and solubility effects. Lehn et al. [164] used this system and applied it for the formation of supramolecular strands, building quasipolymeric associates. Since the cooperative effects are high due to the network formation process, the systems can reach high levels of association and thus higher virtual molecular weights of the associates. Reinhoudt et al. [165] have used this system to assemble calixarenes in a sheet-type fashion (Fig. 47). Thus, two calix [4]arenes bearing either dicyanuric acids or dimelamines were designed, enabling the formation of polymeric aggregates, but prohibiting the formation of cyclic structures. Perfectly aligned strands were observed after deposition of 1 : 1 complexes as a solid. The noncovalent, rod-like structures have a diameter between 3.5 to 5.5 nm and consist of a hydrophobic outer shell and the polar core formed by the hydrogen-bonding networks. Similar systems based on the same or other hydrogen-bonding interactions [166–169] have also led to strand formation via cooperative interactions. Krische et al. have studied the system in detail, putting the focus on the exact match between the supra-

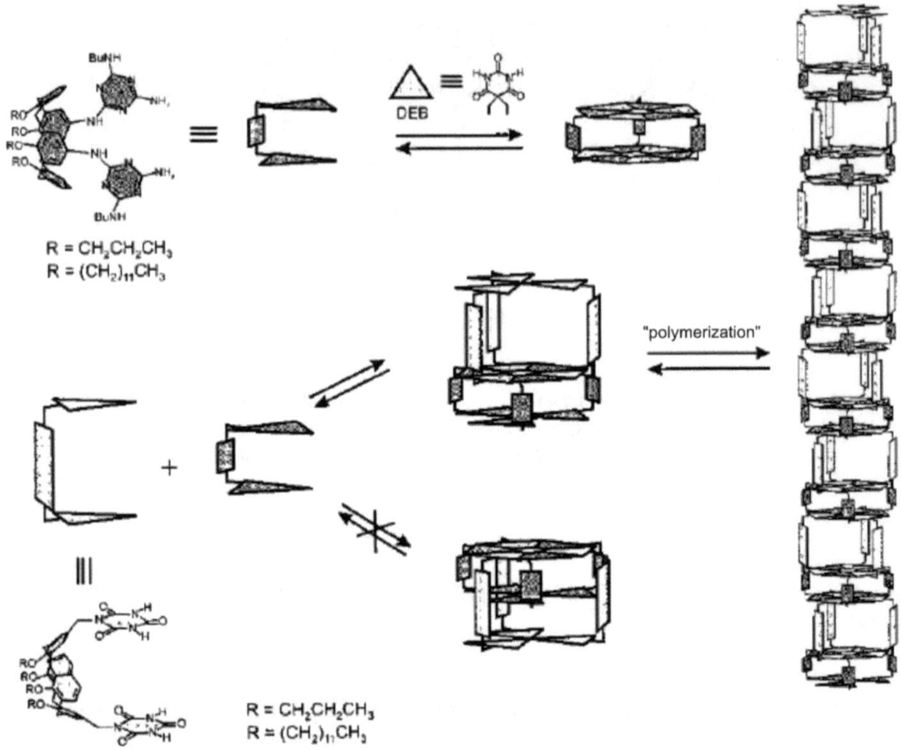

Fig. 47 Assembly of calix[4]arenes via 2,4,6-triamino-triazine/barbiturate interactions. Reprinted with permission from [165]

molecular interaction and the covalent interaction (Fig. 48). The "covalent-casting approach" is described as an approach to fix the supramolecularly ordered assemblies during the ordering process in their predetermined position. This leads to a highly cooperative ordering process, which ends up in the formation of highly organized molecular strands and ribbons. Extending this concept into disc-like aggregates [170] (Fig. 49) leads to discs and platelets, which can hold a photomodulative entity (azobenzene), which causes reversible aggregation and deaggregation phenomena based on the photoisomerization of the azobenzene moiety. Thus, the photoisomerization from the E- to the Z-form was significantly suppressed in the hexamer (rosette) due to the impossibility of the Z-isomer to adapt to the hexameric aggregate due to strong steric interactions. A system consisting of amino-pyrazolones developed by Krische et al. [171] displays a similar formation of cyclic aggregates displaying a network consisting of three associated units.

Apart from those multiple hydrogen-bonding systems described, many others have been investigated, often in relation to association into columns, discs and other networks. Loontjens et al. [172] described networks based on

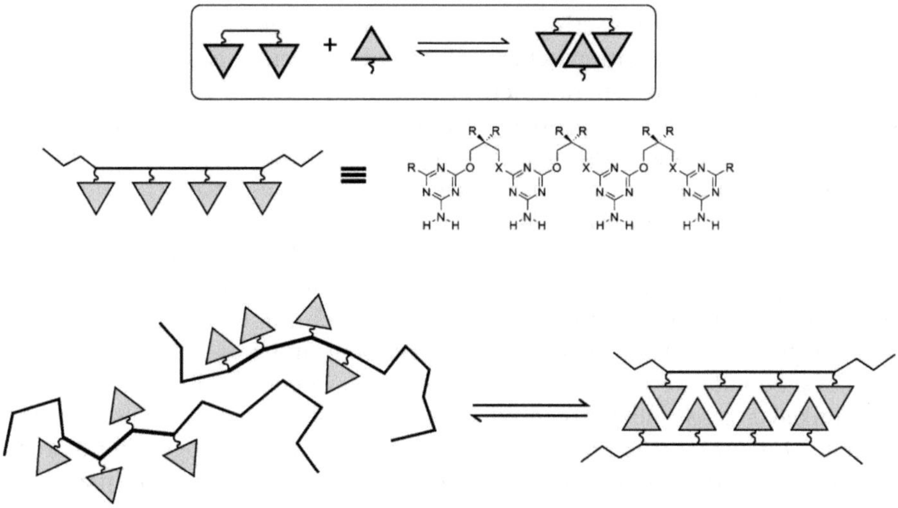

Fig. 48 Assembly of strands by the "covalent casting" approach. Redrawn according to reference [171]

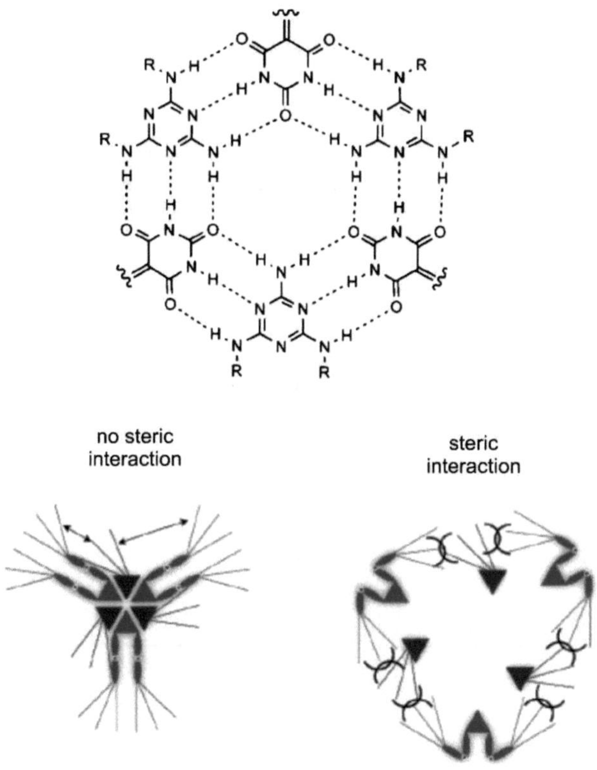

Fig. 49 Assembly of photoisomerizable monomeric units into ordered hexamers. Reprinted with permission from [170]

a 1,3,5-triamino-triazine acting as a central core for the association of imides via a DAD-ADA association scheme. Bivalent molecules bearing two imide anchors were used to generate networks in melt structures with a strongly shear- and temperature-dependent viscosity. Similar approaches using 1,3,5-triamino triazine as a crosslinker to induce crystallization effects and modify rheological properties have been reported with triazine-barbituric acid in the case of poly(propylene) [173, 174].

The association of disc-like building blocks can be achieved using multiple hydrogen bonds on a central, disc-shaped aromatic core (Fig. 50) [175]. Hexagonal columnar structures are formed, which can be stabilized by subsequent polymerization of the terminal acrylate moieties. The central core can subsequently be removed due to the noncovalent bonding between the crosslinked stacks and the central core, leading to nanoporous structures of defined geometry. Thus, the dynamic nature of the hydrogen bond is efficiently used to generate hollow objects. Similar approaches by the same

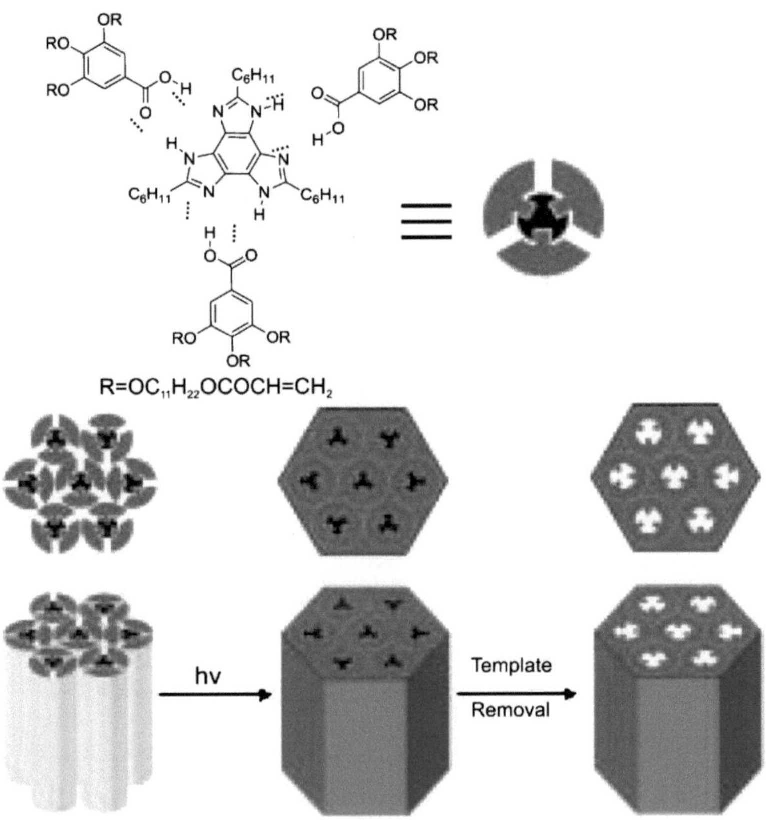

Fig. 50 Formation of nanoporous materials by assembly, subsequent crosslinking and template removal. Reprinted with permission from [175]

authors using guanidinium moieties [176] or 1,3,5-triphenyl-2,4,6-triazine moieties [177] have led to related columnar structures.

Other types of hydrogen-bonding systems using multiple assembly sites have been developed. An important step towards highly self-complementing hydrogen bonds aiming at the formation of defined cyclic structures [178] has been reported by Zimmermann et al. [179] The binding motive shown in

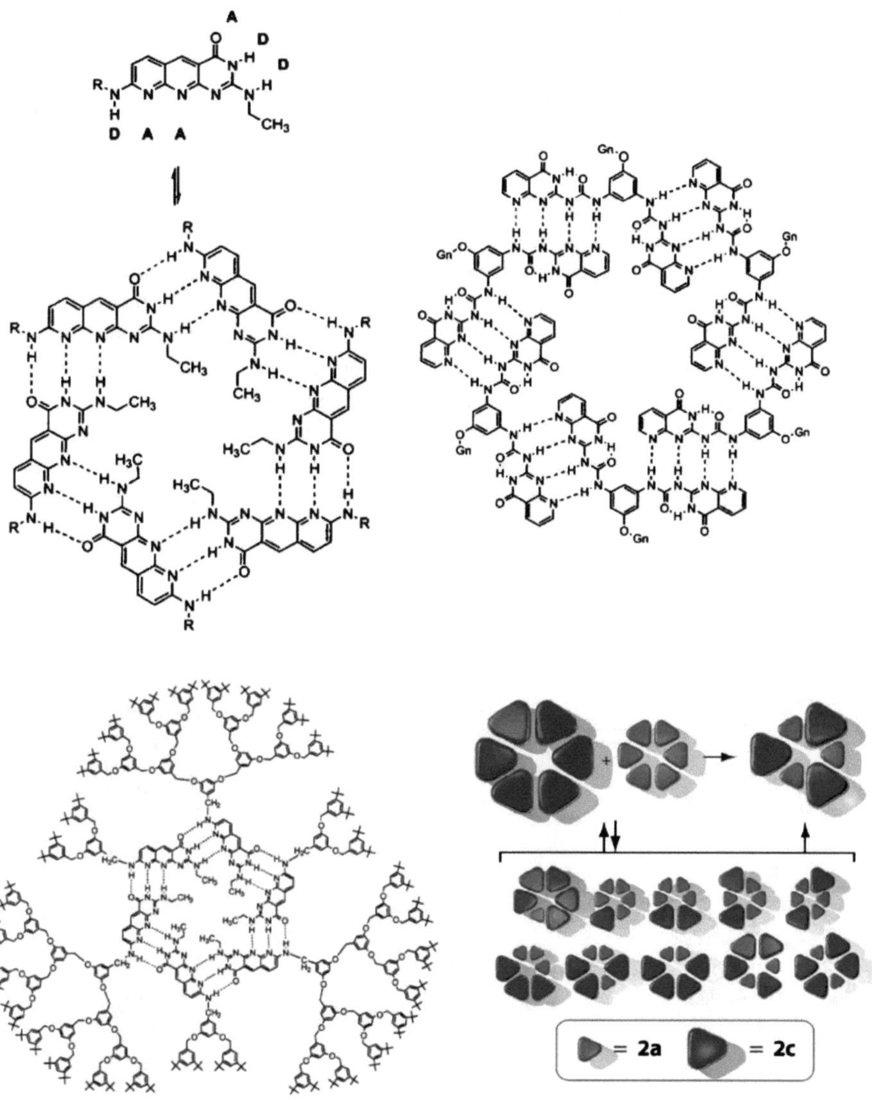

Fig. 51 Assembly of dendritic structures via DDA–AAD hydrogen-bonding motives. Reprinted with permission from [179]

Fig. 51, displaying a DDA-AAD motive, is an extremely strong bonding moiety, with an association constant towards the hexameric associate of more than 10^4 M^{-1} [180]. Most importantly, the stability of this complex is retained not only in nonpolar solvents (such as chloroform and toluene) but also in 15% aqueous THF. The binding motive can be used to assemble dendritic structures bound to the hydrogen-bonding unit. Thus, dendrons up to the 3rd generation were affixed onto the H-bonding motives and assembled into the hexameric aggregates. As expected, the association constant was reduced with increasing steric demand of the dendritic branch. The described system can act as a combinatorial library, where mixtures of different dendrons assemble into aggregates consisting of only separate hexameric species displaying only one type of dendritic branch. Other publications have described dendrimer assembly by either conventional carboxylic acid networks [181–183] or via urea bonds [184, 185].

A related hydrogen-bonding system displaying an AADD self-complementing bonding array was developed by the same authors [186]. An ureidodezapterin was developed, accessible by simple isocyanate addition as in the case of the structurally related ureidopyrimidones. The bond displays an association constant above 10^7 M^{-1} and can be used to assemble polymers, leading to a similar effect as with ureidopyrimidones. Moreover, the ureidodezapterines form 1 : 1 complexes with naphtyridines (DAAD–ADDA interaction) displaying an association constant of > 10^7 M^{-1} (Fig. 52) [187]. Upon affixing those endgroups to poly(styrene) and poly(butylmethacrylate), respectively, a strong association with a significant increase in viscosity was observed.

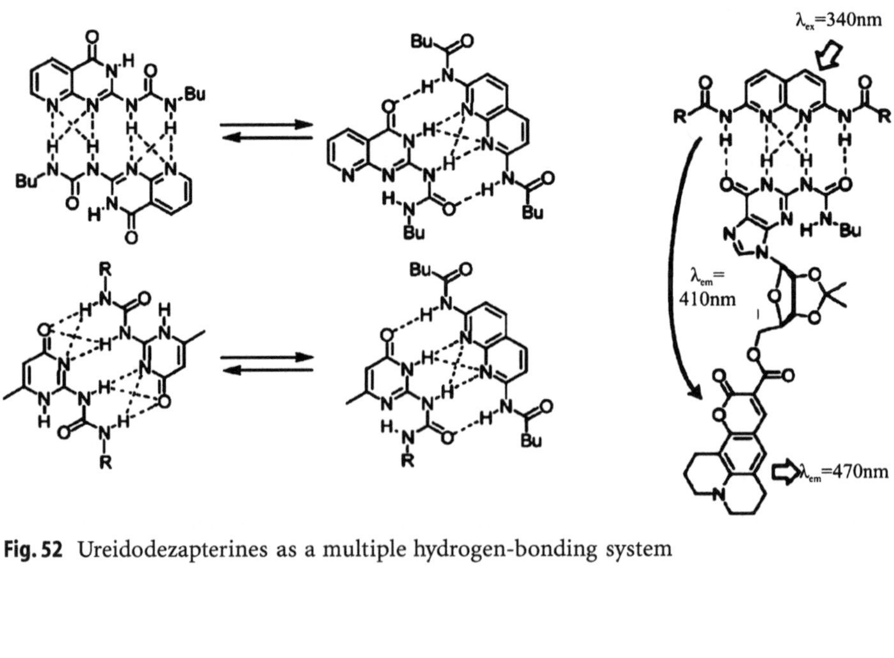

Fig. 52 Ureidodezapterines as a multiple hydrogen-bonding system

3.6 Applications

Hydrogen bonds used as a molecular kit to force the chains of macromolecules into a certain order are definitely not a new principle. Therefore, many aspects of supramolecular ordering, i.e.: in liquid crystals, as crosslinking structures in solid polymers, as an intramolecular force to mediate miscibility effects, and to enable molecular recognition as a principle for applications in sensors and devices are well-known and will not be discussed here. The main novelty of supramolecular polymers lies in two aspects: (1) The possibility to tune a materials properties due to the strength of the hydrogen-bonding interaction. This aspect for the first time allows for an estimation of the principal properties expected in solution or the solid state of a known oligomeric/polymeric material and its properties after incorporation of a specific number of a certain hydrogen bond. (2) Another important aspect concerns the reversible nature of the hydrogen-bonding system, which enables a reversible tuning of the interaction force with temperature, solvents, external fields (shear, electric fields, magnetic fields), and thus the tuning of material properties over a wide range within small external stimuli. These properties lead to materials with definitely new properties and a new engineering design.

Bosman et al. [188] summarized most of the applications related to patenting of the quadrupolar hydrogen bonds, most of all the ureidopyrimidones and the ureidotriazines. Most applications are related to either rheological modifiers (i.e.: tuning the viscosity of a solvent by adding some supramolecular polymer), the material properties exerted in the solid state due to the strong associative force, the phase behavior of the resulting polymers, the properties of inks and polymers for cosmetic use, as well as the use in photoactive and electronic materials, where the supramolecular order induces special optical properties, such as radiation-less photontransfer or electron-transfer processes [189].

Some recent examples including uses as superhydrophobic surfaces [142], medical composites of spun fibers, generating scaffolds for cell attachment [138], thermoplastic elastomers [139], as a supramolecular electrolyte in a dye-sensitized solar cell [190], as a method to align polymer chains [191], or as supramolecular polymer composites [192] have been discussed previously. Still there is ample space to be explored and there definitely will be many more patents and applications in this field.

4 Side-Chain Hydrogen-Bonded Polymers

Similar to supramolecular polymers bearing hydrogen bonds within their main chain, side-chain architectures are possible in a similar manner. Usually, the main backbone of the structure is formed by a polymeric structure rather than a small molecular weight oligomer. Thus, in the case of side-chain architectures, the main chain is already formed by a homo- or block copolymer, which bears an assembled structure of molecules bound to its main chain via hydrogen-bonding interaction, usually in a multiple fashion. The hydrogen-bonding pattern often is formed by "simple" hydrogen bonds, although examples of more complex patterns exist. Thus, usually single-type hydrogen bonds are used, and there are many publications on this topic starting from ~ 1998. The topic has been reviewed during the past years, mostly focussing on the formation of hierarchical structures as well as on their modulative properties [193–196]. A distinct point concerns the influence on miscibility/immiscibility equilibria, most of all the control of polymeric microphases by association/dissociation of the side-chain moieties. If mesogenic units are attached to the polymer, side-chain liquid crystalline materials with special, often highly temperature-sensitive properties are generated. This topic of hydrogen-bonded liquid crystalline materials, also with an emphasis on side-chain liquid crystalline materials, was first reviewed by Kato et al. in 2001 [35] and later in 2006 [36]. Therefore, the interested reader is referred to these reviews.

Another aspect of side-chain hydrogen-bonded polymers is related to the miscibility enhancement between immiscible polymers via (weak) hydrogen-bonding interaction, generating polymer blends. This topic was described in Sect. 3.1, the interested reader is referred to this section within this review.

This section will focus on the microphase behavior of side-chain hydrogen-bonded polymers only. One of the best investigated systems consists of poly(vinylpyridine) with (amphiphilic) phenolic moieties bound to the pyridine units via (ionic) hydrogen bonds. Thus, alkylphenols [such as pentadecyl-phenol (PDP) and nonadecyl phenol (NDP)] form hydrogen bonds to the pendant pyridine moieties (Fig. 53). The structures can be linear, but also crosslinked with bivalent sulfinic acid residues [197]. Lamellar structures, consisting of alternating layers of poly(vinylpyridine) chains, separated by layers of the PDP- or NDP chains are formed [198]. If diblock copolymers (PVP-PS) are used [199], where one block (PVP) acts as a complexing backbone for the PDP, and the other (PS) induces microphase separation, then several structures can be generated, where the microphase separation strongly depends on the temperature (Fig. 54). Thus, below 100 °C layer-within-layer structures are generated, where alternating layers of PS and PVP(PDP) exist, displaying a period of 35 nm. A second microphase ordering, oriented perpendicular to the first one, with a length scale of 4.8 nm exists within the PVP-(PDP) layer. Between 100–150 °C, an order-disorder transition takes place, removing the microphase separation of the lamellae-within-the-lamellae, but keeping the main PS-PVP microphase separation. At temperatures above 150 °C, the lamellae change to cylinders due to the dissociation of the PDP molecules from the main chain

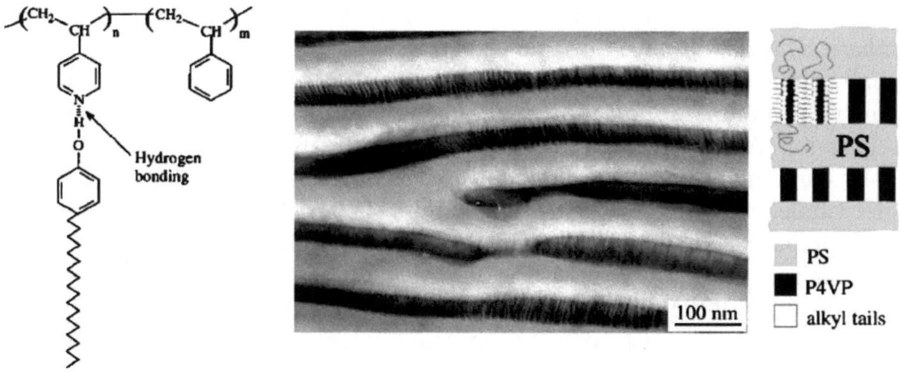

Fig. 53 Formation of lamellar-within-lamellar structures by pentadecylphenol (PDP) of nonadecylphenol (NDP) bound to poly(4-vinylpyridine). Reprinted with permission from [200]

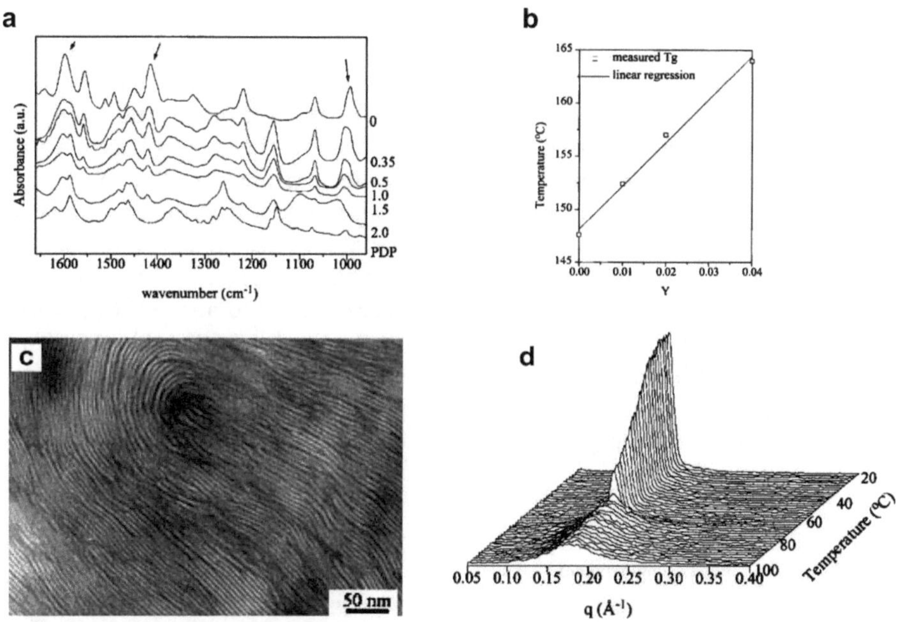

Fig. 54 **a** Hydrogen bond formation monitored by IR-spectroscopy of (PVP(PDP)$_x$). **b** Glass transition temperature as a function of crosslinking density. **c** TEM-micrograph indicating the phase-separated structure. **d** Temperature-dependent SAXS profiles indicating the phase change upon heating. Reprinted with permission from [199]

and their dissolution in the PS and the PVP domain. Thus, the material displays one order/disorder and one order/order transition, as exemplified by temperature-dependent SAXS measurements. More complex structures (i.e.: *lamellar-within-cylindrical*; *spherical-within-lamellar*; *lamellar-within-spherical*) (Fig. 53) have been described using the same system [200]. Many applications to different polymers and networks have been described, relying on this reversible-dissociation phenomenon and the change in microphase separation. An interesting example relying on the use of novolacs as the phenolic component has been demonstrated, using a PVP-*b*-PI diblock copolymer (molecular weights: 2800–30 000 and 21 700–71 000) [201]. The samples were prepared by a solution-blending process and subsequent solvent evaporation and curing at a pressure of 7 MPa up to 190 °C. Figure 55 shows TEM pictures of the final materials, where the weight fraction of the PI-block counts to 0.40. The microphase-separated structure is clearly seen, showing the PI-block in black stripes, and the novolac in between. Reducing the weight fraction of the PI-block to lower amounts leads to worm-like and cylindrical structures. Thus, the morphology of the blend can be adjusted via tailoring of the microphase interaction.

An important factor in achieving functional structures from these microphase-separated polymers lies in the achievement of a high ordering of the microphase domains. Usually this process is achieved via alignment procedures, applying external fields (mostly electrical- and shear-fields) [202]. Several studies, most of them applying shear fields, have been conducted, revealing also the temperature-dependent behavior of the microphases. Thus, the structure and phase equilibria of supramolecular hairy-rodlike polymers consisting of poly(2,5-pyridinium methane sulfonates) have been described in the melt state [203]. Large amplitude oscillatory shear fields were used to achieve the alignment [204, 205]. The shearing was conducted with a rheometer in the oscillatory mode at elevated temperatures (130 and 50 °C, respectively) at 1 Hz shear frequency and a strain amplitude of 50% (Fig. 56). Two different PS-PVP block copolymers and their complexes with either a symmetrical block structure or with only a small PVP block were investigated. The PS cylinders are very-well ordered along the flow in a hexagonal lat-

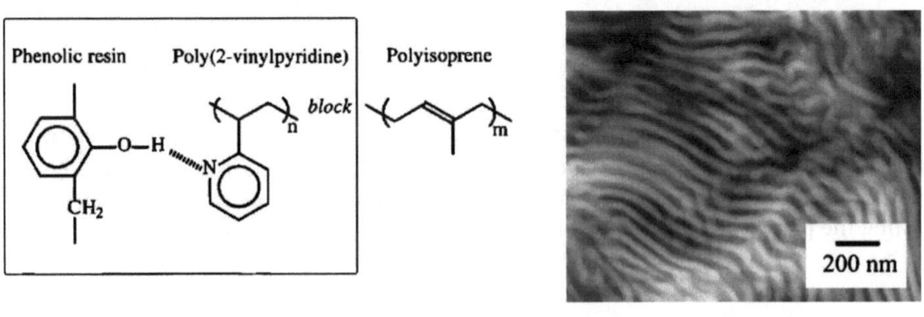

Fig. 55 Chemical structure of complex formation between a phenolic resin and poly(2-vinylpyridine-b-isoprene). TEM-micrograph of the composite, indicating the microphase separation. Reprinted with permission from [201]

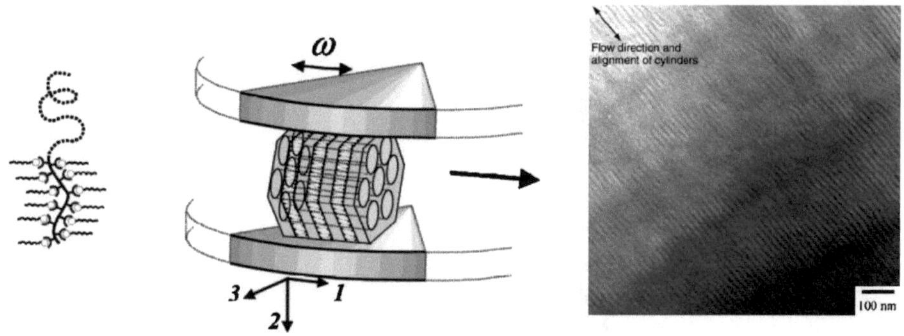

Fig. 56 Shear alignment of a PS-PVP(PDP)x polymer and TEM micrograph of the resulting mixture. Reprinted with permission from [205]

tice, being perfectly aligned such that the (10) plane is parallel to the shear plane. Since the investigated temperature is below the order-disorder temperature of the hexagonal structure, the alignment is claimed to be parallel, rather than perpendicular to the (10) plane. Using this method with PI-PVP diblock copolymers complexed with octyl gallate, together with in-situ SAXS methods reveals order–order transitions between cylinders [206]. Again, the reversible breaking of the hydrogen bond under shear is facilitated, and thus allows an easier alignment of the microphases during this process. The phase behavior of related systems, i.e.: star-[poly(styrene-*b*-2-vinylpyridine)] complexed to dodecyl benzensulfonic acid (DBSA) [207] as well as poly(2,5-pyridinediyl) complexed with camphorsulfonic acid or 5-pentylresocrinol [208] have been reported.

The use of perfectly aligned structures is important when a defined orientation of the microphase-separated structures with respect to a surface is required. One interesting application concerns the engineering of periodic dielectric structures to manipulate the flow of light. In particular, photonic bandgap materials require the presence of three-dimensional periodicity in the range of 100–200 nm to reach the length scale corresponding to $\lambda/2$ of the optical wavelength of light. Since this size range can only be reached via very high molecular weight block copolymers, a different strategy relying on supramolecular polymers was developed by ten Brinke et al. [209] (Fig. 57). Here a P4VP-PS diblock copolymer ($M_{n,PS} = 238\,100$; $M_{n,P4VP} = 49\,500$) with dodecyl benzensulfonic acid (varying from 1.0 to 2.0 mol/P4VP-unit) was used and assembled into 100 mm thin films on quartz glass. Because of the stretching of the chains, very high values of periodicities with a long period of 140 nm were observed, as proven via SAXS and TEM investigations. At specific amounts of the DBSA (1.5), a clear bandgap at 460 nm is observed. Thus, this method allows the generation of functional structures with very high length scales hardly achievable with conventional block copolymers. Similar experiments have been reported for polarized luminescence

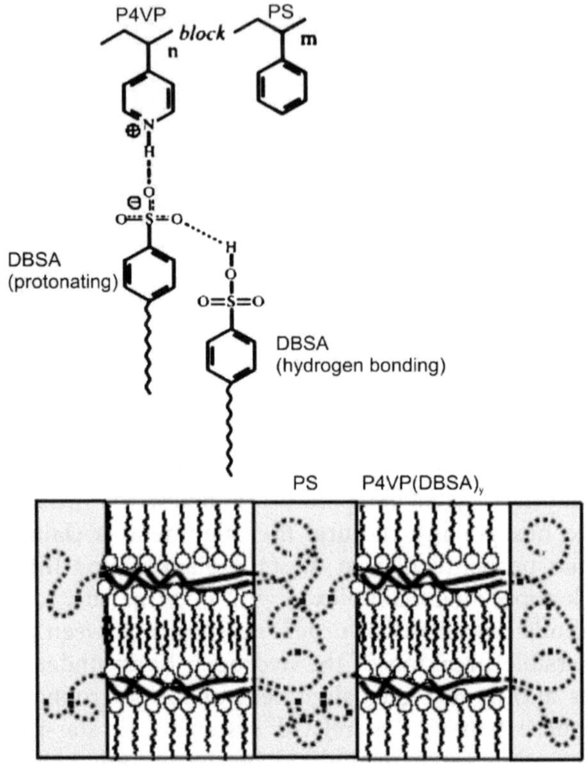

Fig. 57 Generation of large lamellar periodicities via supramolecular ordering of DBSA/P4VP-PS composites. Reprinted with permission from [209]

from self assembled and aligned poly(2,5-pyridinediyl)/camphorsulfonic acid complexes [210].

An interesting approach towards mesoporous materials has been accomplished with PS-b-P4VP diblock copolymers complexed with zinc-dodecylbenzenesulfonate (Fig. 58) [211]. The concept relies on the reversible nature of the hydrogen bond, able to be cleaved and thus extracted after formation of the highly ordered, microphase-separated structure. Thus, after a solution-blending process between the PS-b-P4VP diblock copolymers (238 100 and 49 500 $g\,mol^{-1}$, respectively) and the zinc-dodecylbenzenesulfonate, the structures were annealed forming lamellar phases. Extraction of the zinc-dodecylbenzenesulfonate from the microphase-separated structure leads to mesoporous materials, leaving behind the channels formed by the residual block copolymer. In this example, a coordination bond was included, since it had been observed that purely hydrogen-bonded structures can collapse after extraction [212].

An approach towards fibers made from the hydrogen-bonded supramolecular polymer has been reported [213]. Two PS-b-P4VP diblock copolymers

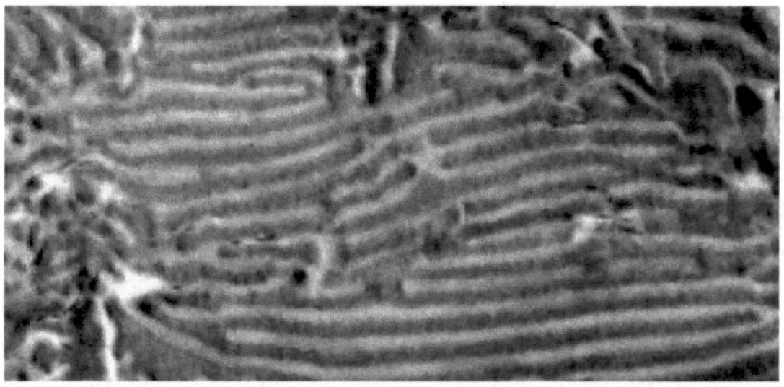

Fig. 58 Generation of mesoporous structures from supramolecular block copolymers. TEM-micrograph after extraction of the zinc-dodecylbenzenesolphonate from a PS-P4VP diblock copolymer mixture. Reprinted with permission from [212]

with different molecular weights (238 100/49 500 and 301 000/19 600 g mol^{-1}) complexed to 3-n-pentadecylphenol were used in an electrospinning process (Fig. 59). This process relies on the formation of a droplet (containing the polymer in solution), which is ejected towards a metallic collector upon applying a strong electric field. Given that the viscosity of the solution is sufficiently high, a continuous jet is formed, generating fibers with entangled polymer chains. This process was applied to solutions of the above-mentioned supramolecular block copolymer, yielding fibers with diameters ranging from 200–400 nm. The resulting materials display a microphase-

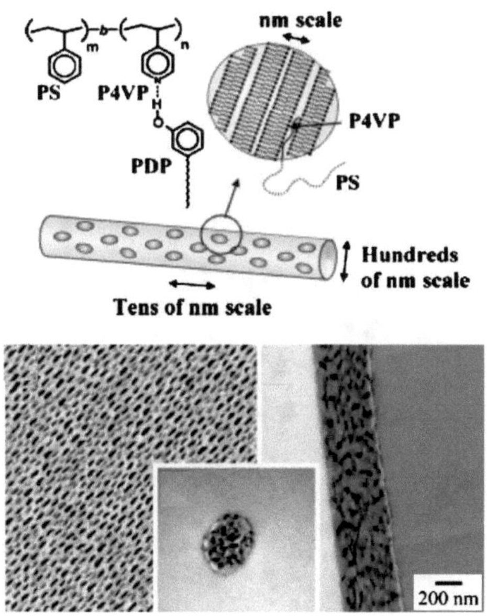

Fig. 59 Formation of fibers by electrospinning from supramolecular PS-b-(P4VP-PDP) polymers. Reprinted with permission from [213]

separated structure, from which the 3-n-pentadecylphenol can be removed by extraction. Again, mesoporous materials are generated, which display a highly ordered character within the now fibrous structure.

If polymers are used, which display electrical or ionic conductivity, the microphase separation will impart a certain level of anisotropy with respect to direction of the conductivity (Fig. 60). Thus, materials with the aforementioned lamellar-within-lamellar phase structures exhibit a strongly tridirectional protonic conductivity [214]. They used PS-b-P4VP block copolymers with stoichiometrically bonded 3-n-pentadecylphenol. The material forms alternating layers with the respective lamellar-within-lamellar structure, with a long period (PS-PVP-layer) of 32 nm, and a short period of 4.1 nm present due to the pentadecyl-phenol layers. As expected, the conductivities were anisotropic for the unaligned material, but different in all three axes for the aligned material, decreasing from the direction parallel to the nanowires, across the pentadecyl layers, and finally to the direction across the PS layers. The hopping rates for the protonic conductivity are largest within the layer, followed by those across the pentadecyl layers, and lowest for the direction across the PS layers.

This concept has been extended using a poly(aniline) as the main conductive polymer (Fig. 61) [215]. Again, a main chain polymer poly(acrylnitrile) able to form ionic hydrogen-bonded complexes (2-acrylamido-2-methyl-

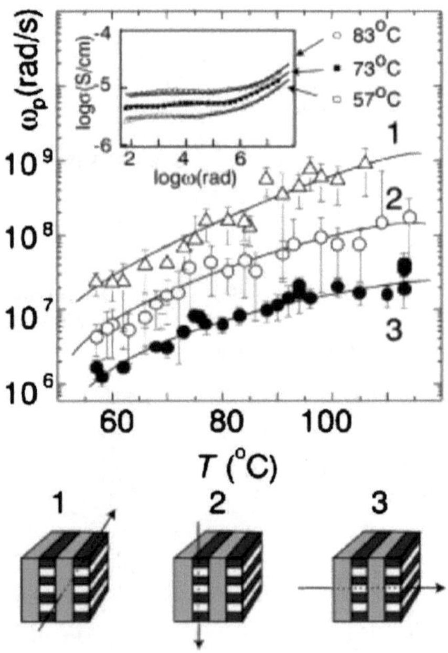

Fig. 60 Generating materials with tridirectional proton-conductivity after phase alignment. Reprinted with permission from [214]

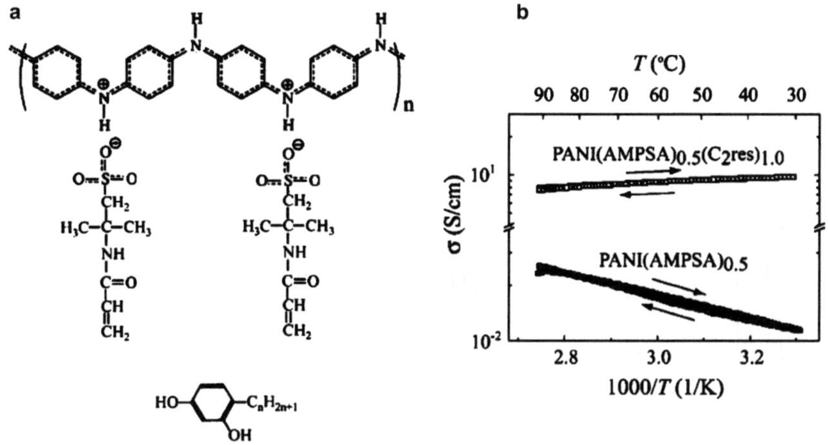

Fig. 61 **a** Conductive supramolecular poly(anilines) with pendant side-chains. **b** Temperature dependent conductivities of the complexes. Reprinted with permission from [215]

1-propanesulfonic acid) and 4-alkyl-resorcinoles was prepared, leading to hydrogen-bonded complexes. SAXS measurements revealed cylindrical assembly, with a tunable long period depending on the type of the 4-alkyl-resorcinol residue. Measuring the conductivity in terms of dependence on the temperature reveals a change from a thermally activated "hopping" mechanism to a "metallic-like" conductivity. An explanation for this behavior lies in a conformational stretching of the PNAI chains upon complexation with specific resorcinoles. Related systems using the PNAI/sulfonic acid system report improved processability [216].

Another, related example of controlling the morphology of a block copolymer via side-chain multiple hydrogen-bonding interactions has been described by Tsao et al. [217] using a diblock copolymer [poly1,4-butadiene-*b*-poly(ethyleneoxide) PB-*b*-PEO]; molecular weight: M_n (PB) = 11 800; M_n (PEO) = 11 000 g mol^{-1} complexed to dodecylbenzene sulfonic acid (DBSA) via the lone-pairs of the PEO side-chain (Fig. 62), thus forming a comb-coil diblock copolymer. In contrast to the expectations, the microphase at lower temperatures ($T < 60\,°C$) is described as PB-cylinders disperged in a PEO/DBSA matrix of lamellar morphology. The PEO-chains are confined within the lamellar mesophase and adopt a folded conformation. At higher temperatures ($T > 60\,°C$) the PB-phase transforms into spheres, disperged within a disordered phase of the PEO/DBSA matrix. This example demonstrates, that strong and significant phase effects can be achieved with relatively weak hydrogen-bonding interactions, devoid of an ionic character.

Significantly stronger bonds, attached to the side-chains of homo- and block copolymers have been used by other authors, relying on multiple bonds rather than single hydrogen bonds as discussed at the beginning of the

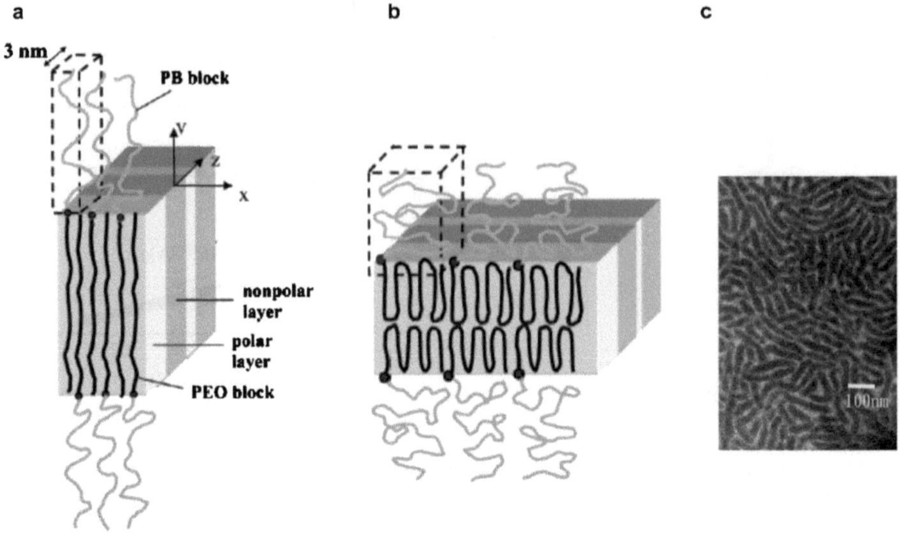

Fig. 62 Formation of complexes between a PB-b-PEO polymer and DBSA. **a** Ordering at $T < 60\,^{\circ}\mathrm{C}$. **b** Ordering at $T > 60\,^{\circ}\mathrm{C}$. **c** TEM picture of the phase-separated structure. Reprinted with permission from [217]

present section. Weck et al. described the use of poly(norbornenes) bearing either 1,6-diamino-pyridine moieties or—alternatively—metal-complexing "pincer" complexes (Fig. 63) [218]. Preparation of the polymers was accomplished via ROMP methodologies of appropriately functionalized norbornene monomers. The resulting polymers show a high density of hydrogen-bonding moieties in their side-chains, forming complexes with the matching *N*-butylthymine. As demonstrated by ^{1}H-NMR investigations, the binding is defined only for the matching hydrogen-bonding interaction. As shown in Fig. 63, the concept can be extended to the construction of block copolymers, presenting an (orthogonal) dual supramolecular scaffold in each specific block [219]. Thus, the binding of specific, matching entities is possible only when a specific interaction (either a hydrogen-bonding interaction or a metal-pincer-complex) is formed. An extension of this concept [220] results in a "universal" polymer backbone, where the two interactions can be used for obtaining various side-chain, supramolecular bonded polymers for further structural studies.

A related concept has been presented by Binder et al. [221] using poly(oxynorbornenes) and different hydrogen-bonding moieties, starting from triple-hydrogen bonds (thymine) to multiple hydrogen-bonding moieties (Hamilton-receptor system) (Fig. 64). The polymers were generated via living ROMP polymerization, using a post-modification reaction, relying on the Sharpless-"click" chemistry. The process involves a 1,3-dipolar cycloaddition reaction between a terminal alkyne and a terminal azide under

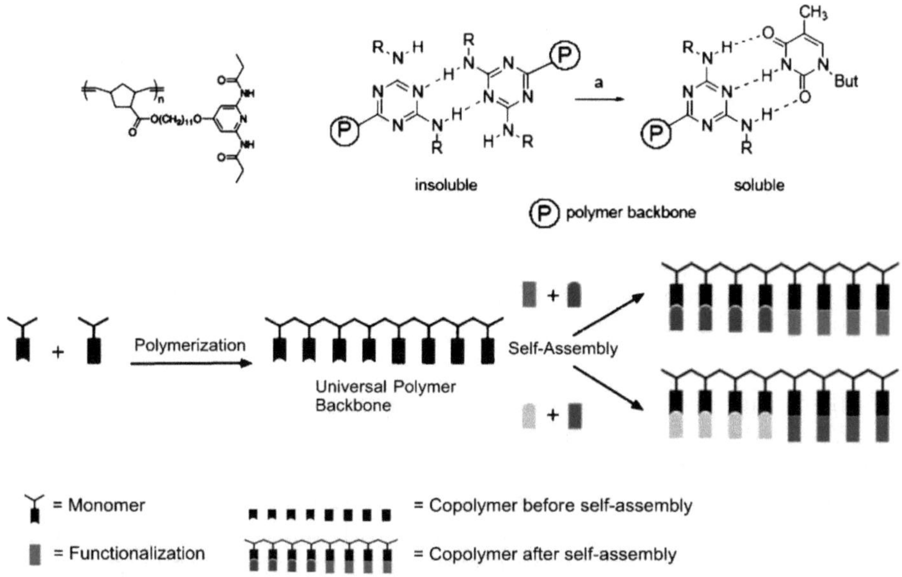

Fig. 63 Generation of block copolymers with orthogonal, matching supramolecular interaction with high densities. Redrawn after citation [218]

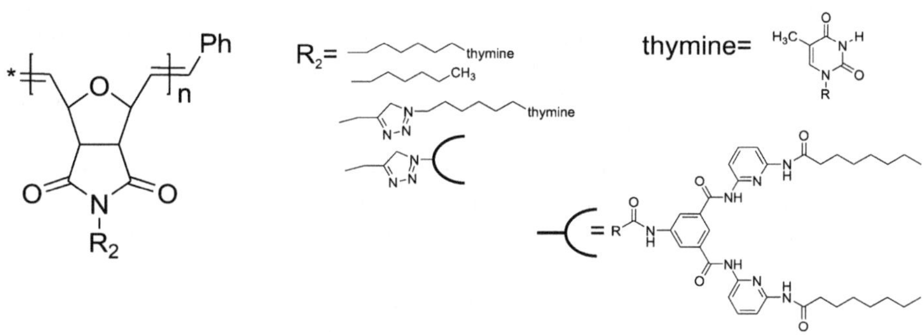

Fig. 64 Poly(oxynorbornenes) with multiple hydrogen-bonding systems

Cu(I)-catalysis. Because of the high efficiency of this reaction, a number of various side-chain modified polymers can be prepared, starting from only a few polymeric scaffolds.

The concept of using two different hydrogen-bonding moieties in each block of a side-chain modified block copolymer has been exploited recently (Fig. 65) [222]. Two different hydrogen-bonding interactions (namely the 2,6-diamino-pyridine/thymine and the Hamilton-receptor/cyanuric acid) were affixed to either of the blocks. The selective complexation of small molecules, bearing the matching hydrogen-bonding interaction, assembled according to their preferential interaction with the respective block structure. Thus,

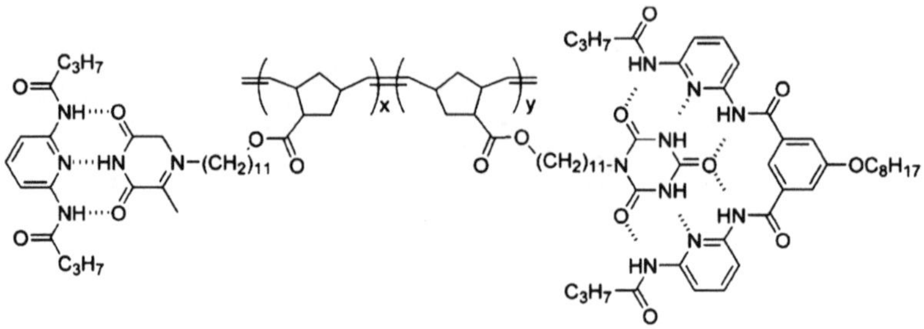

Fig. 65 Block copolymers with two different hydrogen-bonding systems for the investigation of "self"-sorting-phenomena

a "self-sorting" phenomenon was claimed, acting with a similar selectivity as that observed in DNA molecules.

Quadrupolar hydrogen-bonding systems derived from the ureidopyrimidones were presented into 2-ethylhexyl-methacrylates (Fig. 66) [223]. Free radical polymerization led to copolymers with the pendant ureidopyrimidones (1–11 mol%), displaying a molecular weight ranging from 110 000 to 147 000 g mol^{-1}. With increasing amounts of ureidopyrimidone moieties, a linear increase of the glass-transition temperature was observed. The absence of two separate glass-transition temperatures suggests a random copolymerization of the two monomers. Besides creep-testing, also melt-rheological investigations were carried out. The presence of physical crosslinks due to the hydrogen bonds was demonstrated, leading to broadened and increased plateau modulus with increasing content of the hydrogen-bonding moiety.

The self assembly of polymers in the solid state, using polystyrene scaffolds with pendant 2,6-diamino-pyridine (DAP) units has been extensively described by Rotello et al. [224] (Sect. 3.3). Two different polymers were investigated: either a homopolymer, functionalized with the pendant DAP units, or a PS–PS diblock copolymer, in which one block only was functionalized with the DPA units via a p-chloromethylstyrene block (Fig. 67). In order to study the effect exerted by the hydrogen-bonding moieties onto the microphase separation, a series of polymers with different fractions of

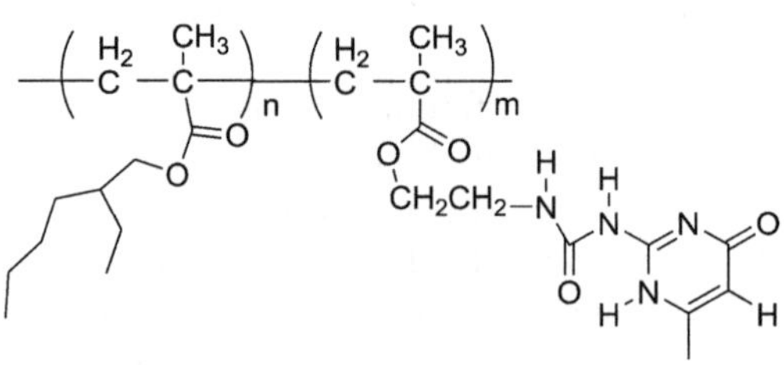

Fig. 66 Copolymers with quadrupolar hydrogen-bonding systems

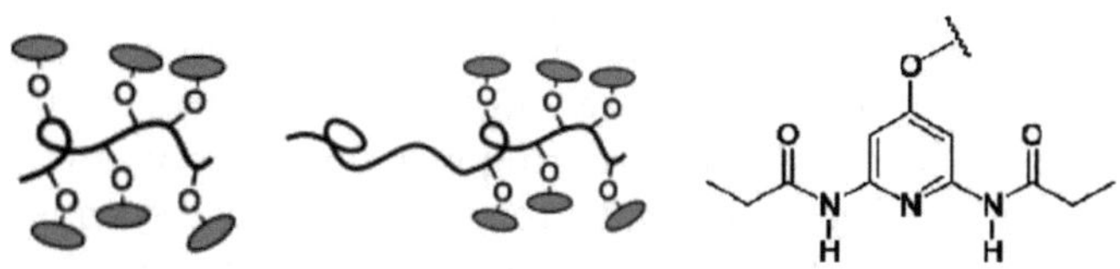

Fig. 67 Homo- and diblock copolymers with pendant 2,6-diamino-pyridine moieties. Redrawn after citation [224]

the DAP block was studied via SAXS methods, analyzing the spacing (D) of the microphases. Since the spacing (D) can be correlated with the Flory–Huggins interaction parameter X, there is a strong correlation of the fraction of the hydrogen-bonding interaction with the spacing of the microphases (D) at a lower fraction of DAP, the polymer is in the strong segregation limit (SSL), whereas at higher fractions of DAP, the weak segregation limit (WSL) is reached. Thus, the number of DAP units can be used as a tool to engineer the microphase of a block copolymer and the respective enthalpic interaction. In a similar approach, the attachment of dendritic structures, relying on thymine/DAP interactions has been studied [225]. A diblock copolymer consisting of a PS-*b*-PS/DAP structure was used as a template for the assembly of dendritic structures bearing the complementing thymine unit, able to attach to the DAP via a triple hydrogen bond. Films were formed from mixtures consisting of the diblock copolymer and the dendrimers, proving the selective enrichment of the dendritic structures within the DAP-containing block. Most importantly, the selective binding was probed by SIMS (dynamic secondary ion mass spectrometry), using a selective, element-specific labeling. Thus, deuterium labels were introduced into the PS/DAP block, and fluorine labels were affixed to the dendritic structures. As proven by SIMS, the elemental distribution showed coincidence between the deuterium and the fluorine labels, proving the sheet structure and the selective binding of the dendrimer to the PS/DAP block via the hydrogen-bonding unit.

Other examples of different polymer systems referring to hydrogels [226] and dendrimers have been described [227], relying mainly on unspecific, but multiple side-chain oriented hydrogen bonds.

5
Hydrogen Bonds on Surfaces

Similar to the formation of hydrogen-bonded oligomeric or polymeric structures in solution, similar ordering processes can take place on surfaces, again using the hydrogen bond as an organizing force. This can lead to the formation of supramolecular order from surfaces into solution, generating "quasipolymeric" structures, being confined to two directions within space. The scientific field stretches strongly into nanoscience, involving not only small molecules, oligomers and macromolecules, but also nanoparticles and the assembly of larger assemblates, relying on similar ordering principles as those in solution. Clearly, similar ordering principles apply as for molecules in solution: again, the specificity and strength of the (hydrogen bonding) interaction will determine the final structure of the aggregates, their virtual degree of polymerization, as well as further assembly from/to the surface. Also, the effects of reversibility of the supramolecular bonds on the final material properties will act in a similar way when applied to surface effects. Therefore, a section reviewing recent achievements of hydrogen bonds, used as molecular scaffolds to assemble polymers, oligomers and nanostructured materials (such as nanoparticles) is included due to its close relationship with supramolecular polymers. Most importantly, AFM- and STM techniques have largely contributed to this field, enabling the molecular study of surfaces and intermolecular surfaces. Several reviews covering the topic of scanning tunnelling microscopy [228] and scanning force microscopy [229] have detailed the possibility to study the exact location, orientation, ordering and the electrical properties of deposited molecules on planar surfaces. Thus, the visualization of individual supramolecular aggregates is possible, revealing a multitude of information of the supramolecular assemblies in solution, enabling one to study assemblies formed via even weak interactions. SFM allowed the visualization of anisotropic hydrogen-bonded networks of single components consisting of 5-(4-dodecyloxybenzylidene)-(1H,3H)2,4,6-pyrimidinetrione/4-amino-2,6-didodecylamino-1,3,5-triazine fibers [230] as well as double components [231]. Moreover, single fibers consisting of isocyanopeptides, bearing pendant L-analyl-D-alanine methyl esters crosslinked via hydrogen-bonding networks have been reported [232]. The molecules can be monitored directly via SFM, revealing an extraordinary stiffness due to the hydrogen-bonded networks between the residues at the side-chain. Additionally, single molecule force spectroscopy has contributed to the understanding of individual molecular interactions [233]. In this example, the deposition of poly(4-vinylpyridine) on quartz, modified with either hydroxyl- or amino groups has been studied. The main interaction between the polymer and the surface is thought to be hydrogen bonding, enabling the individual measurement of the rupture force via single molecule force spectroscopy. Changing the surface from amino groups to hydroxyl groups, leads to a rupture force of 180 pN and 260 pN, respectively. A segmental adsorption is made responsible for this rather large force, exceeding the desorption force of a single hydrogen bond considerably.

Many hydrogen-bonding systems have been used to assemble molecules and objects on surfaces. Apart from the well-known layer-by-layer deposition concept, which relies on single hydrogen-bonding interactions presented in a multiple fashion (usually as a side-chain within a polymer) [234, 235], defined multiple hydrogen-bonding systems have gained interest. Vancso et al. transferred the concept of quadrupolar hydrogen bonds (ureidopyrimidones) onto surfaces via self-assembled monolayers (SAMs) [236]. Thiols bearing the ureidopyrimidone units affixed onto SAMs and monovalent species (bearing dithiol-, trifluoromethyl- and ferrocene-units) were bound via the complementing hydrogen-bonding interaction and studied via AFM measurements (Fig. 68). The attachment of the groups to the surface was found to be very stable, leading to a monostructured layer of the respective functional groups via assembly mediated by the hydrogen bonds. Addition of dipolar aprotic solvents (such as DMSO) or raising the temperature leads to a reversible desorption of the layered structure from the surface. Surprisingly, the complex on the surface was more stable (20 hours) compared to the measured lifetime of the complex in solution (only 170 ms). Thus, surface immobilization can drastically alter the stability of the complexes formed, just leading to enhancement of the supramolecular control. This binding concept has been extended by the same authors towards a "supramolecular polymerization" from the surface using a bivalent chain, bearing two ureidopyrimidone units on either side of a oligo(ethylenglykol)-chain (0.5 ± 1.0 nm) and investigated via AFM-based single molecule force spectroscopy [237] (Fig. 69). Since a bivalent structure is used, a chain extension process, similar to the one known from solution experiments, takes place from the surface-immobilized ureidopyrimidone units. Force displacement curves were measured between an AFM-tip (bearing the immobilized ureidopyrimidone unit) and the surfaces in a solution containing the bivalent "monomer". Besides a rupture force of 172 ± 23 pN, a virtual extension length of 150 nm was observed, corresponding to a virtual degree of polymerization of about 15 units. This example for the first time visualizes the formation of a supramolecular polymer chain, as well as measures the specific interaction force on a molecular level, thus representing a landmark in supramolecular polymer chemistry.

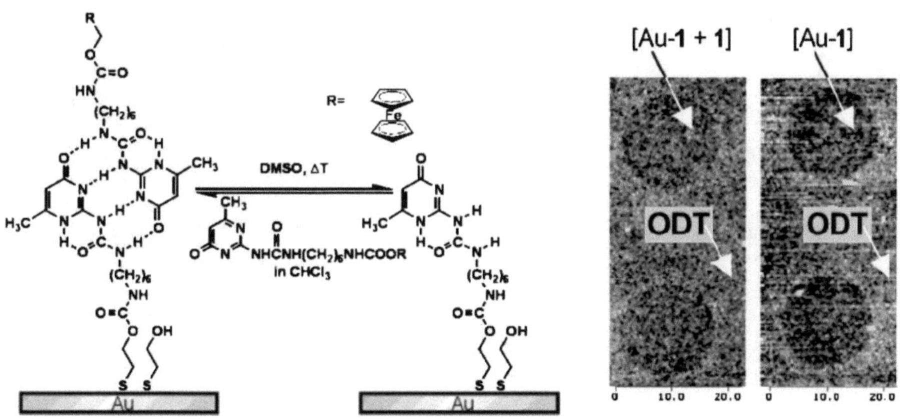

Fig. 68 Binding of ferrocenes to SAM surfaces via quadrupolar hydrogen bonds (ureidopyrimidines). The contact-mode AFM images reveal the enhanced thickness of the layer due to the adsorption process. Reprinted with permission from [236]

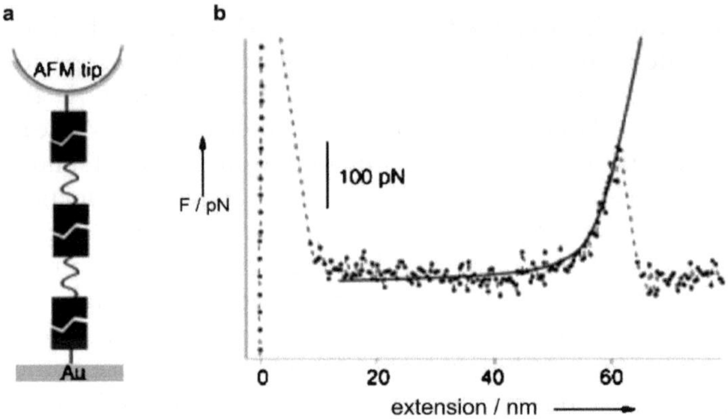

Fig. 69 **a** Formation of supramolecular polymers between an AFM-tip and a SAM via quadrupolar hydrogen bonds. **b** Force-extension curves of the aggregates. Reprinted with permission from [237]

Multiple hydrogen bonds, such as the Hamilton receptor can be used in a similar manner to assemble small functional molecules [238] or even nanoparticles onto SAMs presenting hydrogen-bonding systems. Binder et al. [239] assembled Au-nanoparticles (diameter 5 and 20 nm) onto mixed self-assembled monolayers presenting the Hamilton receptor in a defined density (Fig. 70). 1,3-Dipolar cycloaddition reactions were used to achieve a defined density of the receptors on the surface, enabling molecular densities of the Hamilton receptor ranging from 3 area % to 100 area %. The Au-nanoparticles bearing the corresponding complementing hydrogen-bonding systems were bound in a dense layer on the SAM surface, their deposition den-

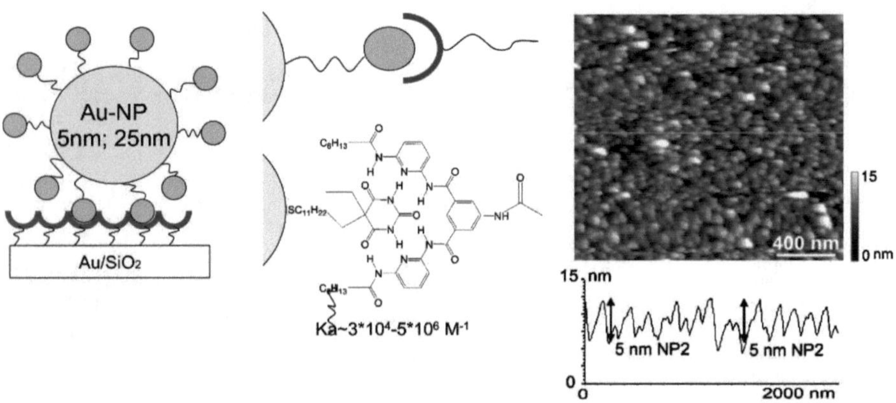

Fig. 70 Binding of Au-nanoparticles to self-assembled monolayer surfaces (SAMs) via specific, multiple hydrogen-bonding interactions. The nanoparticles (5 nm) can be seen as elevations via AFM-imaging methods. Reprinted with permission from [239]

sities depending on the densities of the initially present Hamilton receptor. This concept has also been transferred to polymeric surfaces (Fig. 71) [240]. Thus, diblock copolymers (polyoxynorbornenes) bearing perfluoroalkyl- or

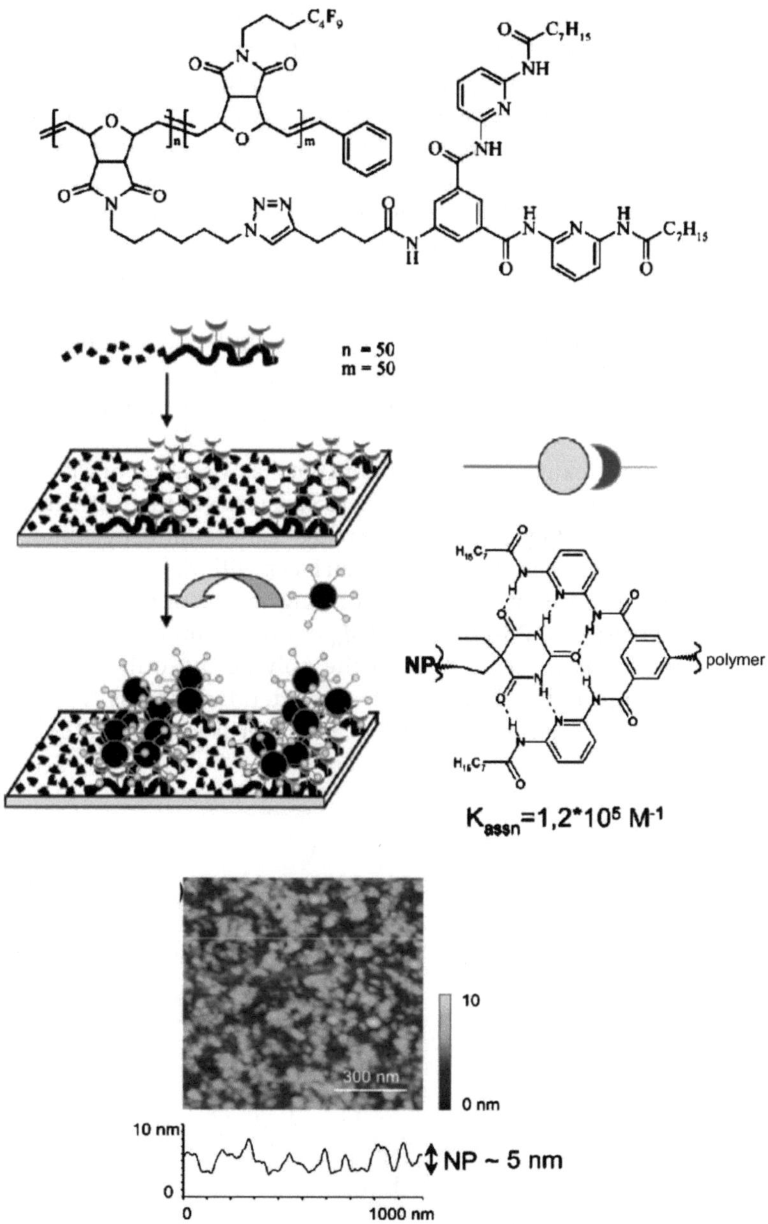

Fig. 71 Assembly of Au-nanoparticles (5 nm) on thin polymeric films made from block copolymers via selective hydrogen-bonding units. Reprinted with permission from [240]

multiple copies of the Hamilton receptor in either block, were deposited as ultrathin films, yielding microphase-separated structures on the surface. Again, Au-nanoparticles bearing the complementing barbituric acid moiety, binding to the Hamilton receptor with a binding constant of $\sim 10^5\ M^{-1}$ were deposited and selectively bound to regions reminiscent of the initial microphase-separated pattern. This concept thus allows the selective deposition of nanoparticles on polymeric films, making use of the microphase separation of block copolymers and their phase pattern on surfaces.

Rotello et al. used a similar system, relying mainly on multiple thymine/2,6-diamino-pyridine interactions (Fig. 72). Small molecules (such as flavines [241], ferrocenes [242]) can be bound via a single, thymine/2,6-diaminopyridine interaction. This can lead to materials with reversible properties, where the redox activity of the surface-bound functional moiety can interact with a current form of the (Au-) surface. The concept has been extended to nanoparticles (Au-NP) [243] and polyhedral oligomeric silsesquioxanes [244]. The presence of the oligomeric silsesquioxanes was proven by XPS measurements, detecting the Si(2p) peak on the surface after deposition.

A very important extension of this concept has been reported using the binding of side-chain-modified polymers to surfaces via the same hydrogen-bonding interaction. Poly(styrenes) bearing 2,6-diamino-pyridine (DAP) units in multiple copies can be bound to surfaces with the matching thymine bond, relying on a trivalent hydrogen-bonding interaction [245]. This concept has been extended to PS-polymers ($M_n = 50\,000\ \mathrm{g\,mol^{-1}}$) bearing a triple copy of the DAP units in order to bind the polymer on only one side, thus generating a polymeric-brush layer of freely extending chains from the surface (Fig. 73) [246]. The so-formed surface forms brushes, where the polymer

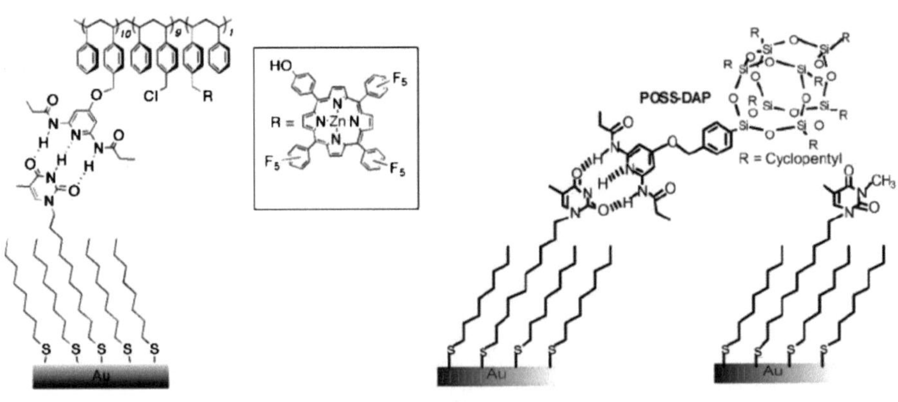

Fig. 72 (*left*) Binding of polymers via selective triple-hydrogen-bonding interactions. (*right*) Selective binding of POSS functionalized with DAP units to SAM surfaces via triple-hydrogen-bonding interactions. Redrawn with permission from [244]

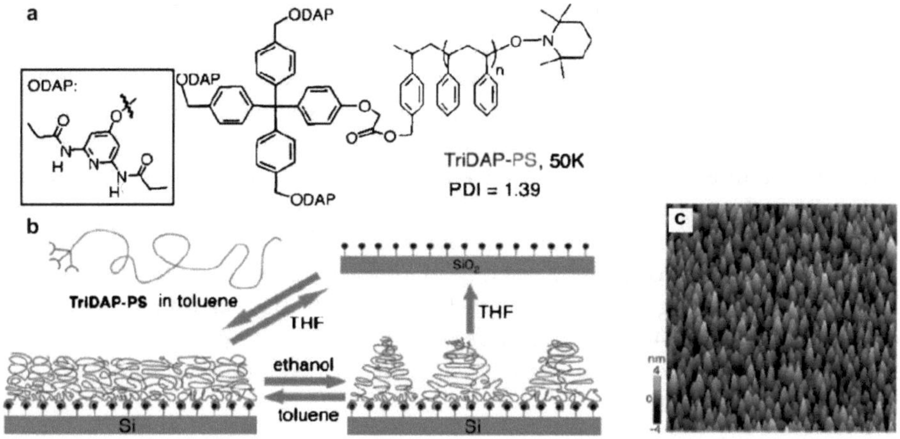

Fig. 73 (**a, b**) Generation of surfaces with responsive properties via selective binding of a PS-polymer bearing triple copies of the DAP units. **c** AFM image showing the corrugated, highly structured surface with the polymer-brushes sticking out of the surface. Reprinted with permission from [246]

chains can be seen "sticking-out" from the surface, while their conformation strongly depends on the solvent used above the polymer layer. Changing the solvent from THF (good solvent for PS) to ethanol (poor solvent for PS), the layer changes its morphology leading to different heights of the layer. Surprisingly, the protic solvent (ethanol) cannot access the hydrogen bond buried underneath the layer structure, being unable to resolve the polymer

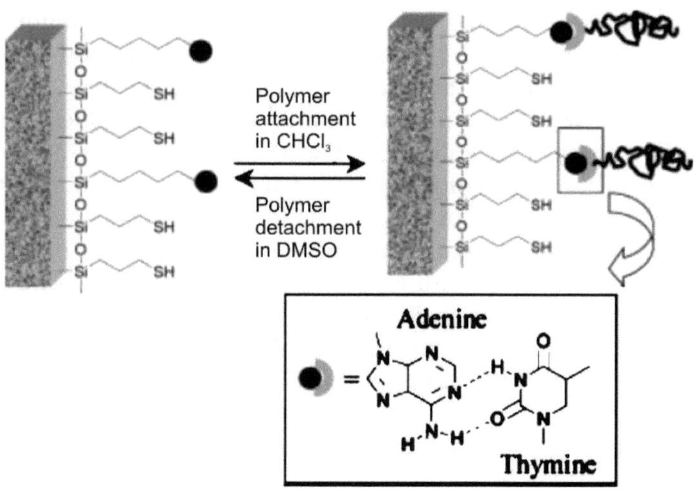

Fig. 74 Selective binding and detachment of a PS-polymer via adenine/thymine interaction to/from a SAM surface. Reprinted with permission from [247]

from the surface. This demonstrates the strong effect of surfaces in relation to solution structures, where the hydrogen bonds would be broken quite efficiently by protic solvents. Furthermore, this approach enables the design of surfaces with strongly modulative properties. A similar approach was reported recently using a PS-polymer ($M_n = 10\,000\ \mathrm{g\,mol^{-1}}$), endfunctionalized with an adenine nucleic acid moiety (Fig. 74) [247]. Again, the attachment of the PS-polymer was achieved via a bivalent hydrogen bond, enabling detachment of the whole polymer by incubation with DMSO. The affixation of the polymer was proven by a variety of physicochemical methods, namely XPS and contact angle measurement. In contrast to the example reported by Rotello et al. [241–244] (see previous paragraph), the PS-polymer can be detached with polar solvents, leading to a (quasi-)reversible adhesion via the supramolecular bond.

6
Conclusions and Future Outlook

Still in its infancy, the field of supramolecular polymer chemistry has definitely found its own area and fixed place within the area of macromolecular and polymer chemistry. Although with a certain delay, the recognition of "designed" intermolecular forces as a tool to direct the ordering and function of macromolecules has now been widely acknowledged and respected. The transfer of principles of "organic" supramolecular chemistry is fully accomplished and used with perfection.

The important contribution of hydrogen bonds to the area of supramolecular polymer chemistry is definitely outstanding, most of all since the potency of hydrogen-bonding systems has been found to be unique in relation to other supramolecular interactions. Thus, the high level of structural diversity of many hydrogen-bonding systems as well as their high level of directionality and specificity in recognition-phenomena is unbeaten in supramolecular chemistry. The realization, that their stability can be tuned over a wide range of binding strength is important for tuning the resulting material properties, ranging from elastomeric to thermoplastic and even highly crosslinked duroplastic structures and networks. On the basis of the thermal reversibility, new materials with highly tunable properties can now be prepared, being able to change their mechanical and optoelectronic properties with very small changes of external stimuli. Thus, the field of hydrogen-bonded polymers forms the basis for stimuli responsive and adaptable materials of the future.

Moreover, the recognition that many aspects of "bulk"-supramolecular polymer chemistry can be transferred to binding and recognition events on surfaces is an area still in its infancy. The binding processes of polymers, nanoparticles or other nanosized objects onto (polymeric, quasipolymeric) surfaces by noncovalent interactions will form a new area in nanoscience and nanotechnology.

The exploitation of the high specificity of the hydrogen-bonding systems, combined with their dynamic features has opened a new branch in polymer science: dynamic materials with self-selection processes. This field, opened up by J.M. Lehn with his "dynamers" is highly prospective for the generation of new materials with properties unachievable with conventional monomers and polymeric materials, relying purely on the covalent bond, instead of the noncovalent, supramolecular interaction.

Many principles exploited during the past years in this field therefore have already found their application in polymeric material science, and definitely will expand in the near future.

Acknowledgements The authors thank the grants FWF 14844 CHE, FWF P18740-B03 and FFG 811021 for financial support.

References

1. Staudinger H (1932) Die Hochmolekularen Organischen Verbindungen. Springer, Berlin, Heidelberg, New York
2. Jiang M, Li M, Xiang M, Zhou H (1999) Adv Polym Sci 146:121
3. Lehn J-M (2002) Polym Int 51:825
4. Stadler R (1993) Kautsch Gummi Kunstst 46:619
5. Lehn J-M (1995) Supramolecular Chemistry. Verlag VCH, Weinheim, p 165
6. Moore JS (1999) Curr Opin Colloid Interf Sci 4:108
7. Amstrong G, Buggy M (2005) J Mater Sci 40:547
8. Ajayaghosh A, George SJ, Schenning APHJ (2005) Top Curr Chem 258:83
9. ten Cate A, Sijbesma RP (2002) Macromol Rapid Commun 23:1094
10. Binder WH (2005) Monatsh Chem 136:1
11. Brunsveld L, Folmer BJB, Meijer EW (2000) MRS-Bulletin 4:49
12. Sherrington DC, Taskinen KA (2001) Chem Soc Rev 30:83
13. Lawrence DS, Jiang T, Levett M (1995) Chem Rev 95:2229
14. Brunsveld L, Folmer BJB, Meijer EW, Sijbesma RP (2001) Chem Rev 101:4071
15. Jeffrey GA, Saenger W (1991) Hydrogen Bonding in Biological Structures. Springer, Berlin, Heidelberg, New York
16. Calhorda MJ (2000) Chem Commun 801
17. Belkova NV, Shubina ES, Epstein LM (2005) Acc Chem Res 38:624
18. Sartorius J, Schneider HJ (1996) Chem Eur J 2:1446
19. Zimmermann SC, Corbin PS (2000) Structure and Bonding, vol 96: Molecular self assembly versus inorganic approaches. Springer, Berlin, Heidelberg, New York, p 63
20. Beijer FH, Sijbesma RP, Vekemans JAJM, Meijer EW, Koojman H, Spek AL (1996) J Org Chem 61:6371
21. Beijer FH, Sijbesma RP, Kooijman H, Spek AL, Meijer EW (1998) J Am Chem Soc 120:6761
22. Prabhakaran P, Puranik VG, Sanjayan GJ (2005) J Org Chem 70:10067
23. Macomber RS (1992) J Chem Edu 5:375

24. Barrans RE, Dougherty DA (1994) Supramol Chem 4:121
25. Bouteiller L, Colombani O, Lortie F, Terech P (2005) J Am Chem Soc 127:8893
26. Arnaud A, Bouteiller L (2004) Langmuir 20:6858
27. Schnell I, Langer B, Söntjens SHM, Sijbesma RP, van Genderern MHP, Spiess HW (2002) Phys Chem Chem Phys 4:3750
28. Spiess HW (2003) Macromol Symp 201:85
29. Söntjens SHM, Sijbesma RP, van Genderen MHP, Meijer EW (2000) J Am Chem Soc 122:7487
30. Loverde SM, Ermoshkin AV, de la Cruz MO (2005) J Polym Sci B Polym Phys 43:796
31. Hue J, Jo WH (2004) Macromolecules 37:3037
32. Chang SK, Hamilton AD (1988) J Am Chem Soc 110:1318
33. Beijer FH, Kooijman H, Spek AL, Sijbesma RP, Meijer EW (1998) Angew Chem Int Ed Engl 37:75
34. Shoji M, Tanaka F (2002) Macromolecules 35:7460
35. Kato T, Mizoshita N, Kanie K (2001) Macromol Rapid Comun 201 22:797
36. Kato T, Mizoshita N, Kishimoto K (2006) Angew Chem Int Ed 45:38
37. Corbin PS, Zimmermann SC (2005) Hydrogen bonded supramolecular polymers: linear and network polymers and self-assembling discotic polymers. CRC press, Boca Raton, Fl, p 153
38. Jiang M, Li M, Xiang M, Zhou H (1999) Adv Polym Sci 146:121
39. Chen D, Jiang M (2005) Acc Chem Res 38:494
40. Ming J, Duan H, Chen D (2003) Macromol Symp 195:165
41. Liu S, Chan CM, Weng LT, Jiang M (2005) J Polym Sci B Polym Phys 43:1924.
42. Liu S, Chan CM, Weng LT, Jiang M (2004) Polymer 45:4945
43. Liu S, Chan CM, Weng LT, Li L, Jiang M (2002) Macromolecules 35:5623
44. Khutoryanskiy VV, Cascone MG, Lazzeri L, Barbani N, Nurkeeva ZS, Mun GA, Dubolazov AV (2004) Polym Int 53:307
45. Orfanou K, Topouza D, Sakellariou G, Pispas S (2003) J Polym Sci A Polym Chem 41:2454
46. Sotiropoulou M, Bokias G, Staikos G (2003) Macromolecules 36:1349
47. Goh HW, Goh SH, Xu GQ (2002) J Polym Sci A Polym Chem 40:4316
48. Huang XD, Goh SH (2000) Macromolecules 33:8894
49. Quyang J, Hong GS, Li Y (2001) Chem Phys Lett 347:344
50. Yi JZ, Goh SH (2001) Polymer 42:9313
51. Qiu F, Chen S, Ping Z (2005) Magn Reson Chem 43:411
52. Zhang SH, Jin X, Painter PC, Runt J (2003) Macromolecules 36:5710
53. Liu S, Zhu H, Zhao H, Jiang M (2000) Polymer 42:151
54. Goh SH, Liu Y, Lee SY, Huan CHA (1999) Macromolecules 32:8595
55. Abdellaoui N, Djadoun S (2005) J Appl Polym Sci 98:658
56. Kutoryanskaya O, Khutoryanskij VV, Pethrick RA (2005) Macromol Chem Phys 205:1479
57. Khutoryanskij VV, Cascone MG, Lazzeri L, Barbani M, Nurkeeva ZS, Mun GA, Dubolazov AV (2004) Polym Int 53:307
58. Li X, Goh SH, Lai YH, Wee ATS (2001) Polymer 42:5463
59. Yi JZ, Goh SH (2002) Polymer 43:4515
60. Ni Y, Zheng S (2005) Polymer 46:5828
61. Brinkmann-Rengel S, Abetz V, Stadler R, Thomas EL (1999) Kautsch Gummi Kunstst 52:806
62. Bica CID, Burchard W, Stadler R (1996) Macromol Chem Phys 197:3407

63. Schadebrodt J, Ludwig S, Abetz V, Stadler R (1999) Kautsch Gummi Kunstst 52:555
64. Hellmann J, Hilger C, Stadler R (1992) Polym Adv Technol 5:763
65. Hilger C, Stadler R (1991) Polymer 32:3244
66. Müller M, Dardin A, Seidel U, Balsamo V, Ivan B, Spiess HW, Stadler R (1996) Macromolecules 29:2577
67. Schadebrodt J, Ludwig S, Abetz V, Stadler R (1999) Kautsch Gummi Kunstst 52:555
68. Hellmann J, Hilger C, Stadler R (1992) Polym Adv Technol 5:763
69. Müller M, Seidel U, Stadler R (1995) Polymer 36, 16:3143
70. Tang H, Sun J, Zhou X, Fu P, Xie P, Zhang R (2003) Macromol Chem Phys 204:155
71. Li BS, Cheuk KKL, Ling L, Chen J, Xiao X, Bai C, Tang BZ (2003) Macromolecules 36:77
72. Cheuk KKL, Lam JWY, Lai LM, Dong Y, Tang BZ (2003) Macromolecules 36:9752
73. Kim C, Lee SJ, Lee IH, Kim KT (2003) Chem Mater 15:3638
74. Rakotondradany F, Whitehead MA, Lebuis AM, Sleiman HF (2003) Chem Eur J 9:4771
75. Zeng F, Zimmermann SC, Kolotuchin SV, Reichert DEC, Ma Y (2002) Tetrahedron 58:825
76. Duweltz D, Laupetre F, Abed S, Bouteiller L, Boileau S (2003) Polymer 44:2295
77. Abed S, Boileau S, Bouteiller L (2001) Polymer 42:8613
78. Abed S, Boileau S, Bouteiller L (2000) Macromolecules 33:8479
79. Bazzi HS, Sleiman HF (2002) Macromolecules 26:9617
80. Sivakova S, Rowan SJ (2005) Chem Soc Rev 34:9
81. Fouquey C, Lehn JM, Levelut AM (1990) Adv Mater 5:254
82. Kotera M, Lehn JM, Vigneron JP (1995) Tetrahedron 51:1953
83. Lange RFM, Meijer EW (1995) Macromolecules 28:782
84. Asunama H, Ban T, Gotho S, Hishiya T, Komiyama M (1998) Macromolecules 31:371
85. Asunama H, Ban T, Gotho S, Hishiya T, Komiyama M (1998) Supramol Sci 5:405
86. Würthner F, Thalacker C, Sautter A (1999) Adv Mater 11:754
87. Yamauchi K, Lizotte JR, Long TE (2002) Macromolecules 35:8745
88. Yamauchi K, Long TE (2002) J Am Chem Soc 124:8599
89. Yamauchi K, Long TE (2001) Polym Mater Sci Eng 85:465
90. Chino K, Ashimura M (2001) Macromolecules 34:9201
91. Mather BD, Lizotte JR, Long TE (2004) Macromolecules 37:9331
92. Kunz MJ, Hayn G, Saf R, Binder WH (2004) J Polym Sci A Polym Chem 42:661
93. Binder WH, Kunz MJ, Ingolic E (2004) J Polym Sci A Polym Chem 42:162
94. Yang J, Ding S, Radosz M, Shen Y (2004) Macromolecules 37:1728
95. Ding S, Yang J, Radosz M, Shen Y (2003) J Polym Sci A Polym Chem 42:22
96. Dahman Y, Puskas JE, Margaritis A (2003) Macromolecules 36:2198
97. Deans R, Rotello VM (1997) J Org Chem 62:4528
98. Deans R, Ilhan F, Rotello VM (1999) Macromolecules 32:4956
99. Ilhan F, Gray M, Rotello VM (2001) Macromolecules 34:2597
100. Cooke G, Rotello VM (2002) Chem Soc Rev 31:275
101. Carroll JB, Waddon AJ, Nakade H, Rotello VM (2003) Macromolecules 36:6289
102. Uzun O, Frankamp BL, Rotello VM (2002) Polym Prepr 43:419
103. Bourgel C, Boyd ASF, Cooke G, Augier de Cremiers H, Duclairoi FMA, Rotello VM (2001) Chem Commun 1955
104. Thibault RJ, Hotchkiss PJ, Gray M, Rotello VM (2003) J Am Chem Soc 125:11249
105. Drechsler U, Thibault RJ, Rotello VM (2002) Macromolecules 35:9621
106. Das K, Nakade H, Penelle J, Rotello VM (2004) Macromolecules 37:310

107. Sivakova S, Wu J, Campo SJ, Mather PT, Rowan SJ (2006) Chem Eur J 12:446
108. Sivakova S, Rowan SJ (2003) Chem Commun 2428
109. Shimizu T, Iwaura R, Masuda M, Hanada T, Yase K (2001) J Am Chem Soc 123:5947
110. Sivakova A, Bohnsack DA, Mackay ME, Suwanmala P, Rowan SJ (2005) J Am Chem Soc 127:18202
111. Noro A, Nagata Y, Tsukamoto M, Hayakawa Y, Takano A, Matsushita Y (2005) Biomacromolecules 6:2328
112. Khan A, Haddleton DM, Hannon MJ, Kukulj D, Marsh A (1999) Macromolecules 32:6560
113. Marsh A, Khan A, Haddleton DM, Hannon MJ (1999) Macromolecules 32:8725
114. Li Z, Ding J, Day M, Tao Y (2006) Macromolecules 39:2629
115. Öjelund K, Loontjens P, Palmans A, Maurer F (2003) Macromol Chem Phys 204:52
116. Scmuck C, Wienand W (2001) Angew Chem Int Ed 40:4363
117. Brunsveld L, Vekemans JAJM, Hirschberg JHKK, Sijbesma RP, Sijbesma RP, Meijer EW (2002) Proc Nat Acad Sci USA 99:4977
118. Lange RFM, Gurp M, Meijer EW (1999) J Polym Sci A Polym Chem 37:3657
119. Keizer HM, van Kessel R, Sijbesma RP, Meijer EW (2003) Polymer 44:5505
120. Bosman AW, Brunsveld L, Folmer BJB, Sijbesma RP, Meijer EW (2003) Macromol Symp 201:143
121. Sijbesma RP, Beijer FH, Brunsveld L, Folmer BJ, Lange RF, Lowe JK, Meijer EW (1997) Science 278:1577
122. Jacobson H, Stockmayer WH (1950) J Chem Phys 18:1600
123. ten Cate AT, Kooijman H, Spek AL, Sijbesma RP, Meijer EW (2004) J Am Chem Soc 126:3801
124. Söntjens SHM, Sijbesma RP, van Genderen MHP, Meijer EW (2001) Macromolecules 34:3815
125. ten Cate AT, Dankers PYW, Kooijman H, Spek AL, Sijbesma RP, Meijer EW (2003) J Am Chem Soc 125:6860
126. Folmer BJB, Cavini E, Sijbesma RP, Meijer EW (1998) Chem Commun 1847
127. Takeshita M, Hayashi M, Kadota S, Mohammed KH, Yamato T (2005) Chem Commun 761
128. Dudek SP, Pouderoijen M, Abbel R, Schenning APHJ, Meijer EW (2005) J Am Chem Soc 127:11763
129. El-ghayoury A, Schenning APHJ, van Hal P, van Duren KJ, Janssen RAJ, Meijer EW (2001) Angew Chem Int Ed 40:3660
130. El-ghayoury A, Peters E, Schenning APHJ, Meijer EW (2000) Chem Commun 1969
131. Folmer BJB, Sijbesma RP, Versteegen RM, van der Rijt JAJ, Meijer EW (2000) Adv Mater 12:874
132. Yamauchi K, Lizotte JR, Long TE (2003) Macromolecules 36:1083
133. KyHirschberg JHK, Beijer FH, van Aert HA, Magusin PCMM, Sijbesma RP, Meijer EW (1999) Macromolecules 32:2696
134. Yamauchi K, Kanomata A, Inoue T, Long TE (2004) Macromolecules 37:3519
135. Rieth LR, Eaton RF, Coates GW (2001) Angew Chem Int Ed 113:221
136. Hofmeier H, Hoogenboom R, Wouters MEL, Schubert US (2005) J Am Chem Soc 127:2913
137. Hofmeier H, El-ghayoury A, Schenning APHJ, Schubert US (2004) Chem Commun 318
138. Dankers PYW, Harmsen MC, Brouwer LA, van Luyn MJA, Meijer EW (2005) Nat Mater 7:568

139. Ligthart GBWL, Ohkawa H, Sijbesma RP, Meijer EW (2006) J Am Chem Soc 127:810
140. Yamauchi K, Lizotte JR, Hercules DM, Vergne MJ, Long TE (2002) J Am Chem Soc 124:8599
141. Keizer HM, Sijbesma RP, Jansen JFGA, Pasternack G, Meijer EW (2003) Macromolecules 36:5602
142. Han JT, Lee DH, Ryu CH, Cho K (2004) J Am Chem Soc 126:4796
143. Feng L, Li S, Li Y, Li H, Zhang L, Zhai J, Song Y, Lui B, Jaing L, Zhu D (2002) Adv Mater 14:1857
144. Marmur A (2004) Langmuir 20:3517
145. KyHirschberg JHK, Brunsveld L, Ramzi A, Vekemans JAJM, Sijbesma RP, Meier EW (2000) Nature 407:167
146. KyHirschberg JHK, Ramzi A, Sijbesma RP, Meijer EW (2003) Macromolecules 36:1429
147. Öjelund K, Loontjens T, Steeman P, Palmans A, Maurer F (2003) Macromol Chem Phys 204:52
148. Tecilla P, Hamilton AD (1990) J Chem Soc Chem Commun 1232
149. Chang SK, van Engen D, Hamilton AD (1991) J Am Chem Soc 113:7640
150. Berl V, Schmutz M, Krische MJ, Khoury RG, Lehn JM (2002) Chem Eur J 8:1227
151. Kolomiets E, Buhler E, Candau SJ, Lehn JM (2006) Macromolecules 39:1173
152. Berl V, Krische MJ, Huc I, Lehn JM, Schmutz M (2000) Chem Eur J 6:1938
153. Berl V, Huc I, Khoury RG, Krische MJ, Lehn JM (2000) Nature 407:720
154. Zhuang J, Zhou W, Li X, Li Y, Wang N, He X, Liu H, Li Y, Jiang L, Huang C, Cui S, Wang S, Zhu D (2005) Tetrahedron 61:8686
155. Kanazawa H, Higuchi M, Yamamoto K (2006) Macromolecules 39:138
156. Kolomiets E, Lehn JM (2005) Chem Commun 1519
157. Binder WH, Bernstorff S, Kluger C, Petraru L, Kunz MJ (2005) Adv Mater 17:2824
158. Kunz MJ, Hayn G, Saf R, Binder WH (2004) J Polym Sci A Polym Chem 42:661
159. Binder WH, Kunz MJ, Kluger C, Hayn G, Saf R (2004) J Polym Sci A Polym Chem 37:1749
160. Binder WH, Kunz MJ, Ingolic E (2004) J Polym Sci 42:162
161. Binder WH, Petraru L, Roth T, Groh PW, Palfi V, Keki S, Ivan B (2006) Adv Funct Mater ASAP, in press
162. Yang X, Hua F, Yamato K, Ruckenstein E, Gong B, Kim W, Ryu CY (2004) Angew Chem 43:6471
163. Whitesides GM, Simanek EE, Mathias JP, Seto CT, Chin DN, Mammen M, Gordon DM (1995) Acc Chem Res 28:37
164. Lehn JM, Mascal M, DeCian A, Fischer J (1990) J Chem Soc Chem Commun 479
165. Klok HA, Jolliffe KA, Schauer CL, Prins LJ, Spatz JP, Möller M, Timmerman P, Reinhoudt DN (1999) J Am Chem Soc 121:7154
166. Mascal M, Hansen J, Fallon PS, Blake AJ, Heywood BR, Moore MH, Turkenburg JP (1999) Chem Eur J 5:381
167. Gong H, Krische MJ (2005) J Am Chem Soc 127:1719
168. Yagai S, Iwashima T, Karatsu T, Kitamura A (2004) Chem Commun 1114
169. Yagai S, Higashi M, Karatsu T, Kitamura A (2005) Chem Mater 17:4392
170. Yagai S, Nakajima T, Karatsu T, Saitow K, Kitamura A (2004) J Am Chem Soc 126:11500
171. Gong H, Krische MJ (2005) Angew Chem Int Ed 44:7069
172. Loontjens T, Put J, Coussens B, Lange R, Palmen J, Sleijpen T, Plum B (2001) Macromol Symp 174:357

173. Fahrländer M, Fuchs R, Mühlhaupt C, Friedrich C (2003) Macromolecules 36:3749
174. Fuchs K, Bauer T, Thormann C, Wang C, Friedrich R, Mühlhaupt R (1999) Macromolecules 32:8404
175. Lee H-K, Lee H, Ko YH, Chang YJ, Oh NK, Zin WC, Kim K (2001) Angew Chem Int Ed 40(14):2669
176. Dongwoo K, Sangyong J, Lee HK, Baek K, Oh N-K, Zin W-C, Kim K (2005) Chem Comm 44:5509
177. Lee H, Kim D, Lee H-K, Qiu W, Oh N-K, Zin W-C, Kim K (2004) Tetrahedron Lett 45:1019
178. Corbin PS, Lawless LJ, Li Z, Ma Y, Witmer MJ, Zimmermann SC (2002) Proc Nat Acad Sci USA 99:5099
179. Ma Y, Kolotouchin SV, Zimmermann SC (2002) J Am Chem Soc 124:13757
180. Kolotouchin SV, Zimmermann SC (1998) J Am Chem Soc 120:9092
181. Zimmermann SC, Zeng F, Reichert DEC, Kolotouchin SV (1996) Science 271:1095
182. Thiyagarajan P, Zeng F, Ku CY, Zimmermann SC (1997) J Mater Chem 7:1221
183. Zeng F, Zimmermann SC, Kolotouchin SV, Reichert DEC, Ma Y (2002) Tetrahedron 58:825
184. Gillies ER, Frechet JMJ (2004) J Org Chem 69:46
185. Versteegen RM, van Beek DJM, Sijbesma RP, Vlassopoulos D, Fytas G, Meijer EW (2005) J Am Chem Soc 127:13862
186. Corbin PS, Zimmermann SC (1998) J Am Chem Soc 120:9710
187. Park T, Zimmermann SC, Nakashima S (2005) J Am Chem Soc 127:6520
188. Bosman AW, Sijbesma RP, Meijer EW (2004) Mater Today 4:34
189. Dudek SP, Pouderoijen M, Abbel R, Schenning APHJ, Meijer EW (2005) J Am Chem Soc 127:11763
190. Kim YJ, Kim JH, Kang MS, Lee MJ, Won J, Lee JC, Kang YS (2004) Adv Mater 16:1753
191. Knaapila M, Lockhart E, Monkman AP, Ikkala O (2003) Patent WO03082948
192. Fomperie L, Bouteiller L, Colombani O (2003) US Patent 2003092838
193. Ikkala O, ten Brinke G (2004) Chem Commun 2131
194. ten Brinke G, Ikkala O (2004) Chem Rec 4:219
195. ten Brinke G, Ikkala O (2003) Macromol Symp 203:103
196. Ikkala O, ten Brinke G (2002) Science 295:2407
197. Shibata M, Kimura Y, Yaginuma D (2004) Polymer 45:7571
198. Ruokolainen J, Tanner J, Ikkala O, ten Brinke G, Thomas EL (1998) Macromolecules 31:3532
199. Ruokolainen J, Mäkinen R, Torkkeli M, Mäkelä T, Serimaa R, ten Brinke G, Ikkala O (1998) Science 280:557
200. Ruokolainen J, ten Brinke G, Ikkala O (1999) Adv Mater 11:777
201. Kosonen H, Ruokolainen J, Nyholm P, Ikkala O (2001) Macromolecules 34:3046
202. Li M, Coenjarts CA, Ober CK (2005) Adv Polym Sci 190:183
203. Knaapila M, Stepanyan R, Horsburgh LE, Monkman AP, Serimaa R, Ikkala O, Subbotin A, Torkkeli M, ten Brinke G (2003) J Phys Chem B 107:14199
204. van Ekenstein GA, Polushkin E, Nijland H, Ikkala O, ten Brinke G (2003) Macromolecules 36:3684
205. Ruotsolainen T, Torkkeli M, Serimaa R, Mäkelä T, Mäki-Intto R, Ruokolainen J, ten Brinke G, Ikkala O (2003) Macromolecules 36:9437
206. Polushkin E, Bondzic S, de Wit J, van Ekenstein GA, Dolbnya I, Bras W, Ikkala O, ten Brinke G (2005) Macromolecules 38:1804
207. Nandan B, Lee CH, Chen HL, Chen WC (2005) Macromolecules 38:10117

208. Knaapila M, Torkkeli M, Palsson LO, Horsburgh LE, Jokela K, Dolbnya I, Bras W, Serimaa R, ten Brinke G, Monkman AP, Ikkala O (2002) Mater Res Soc Symp Proc 725:237
209. Kosonen H, Valkma S, Ruokolainen J, Torkkeli M, Serimaa R, ten Brinek G, Ikkala O (2003) Eur Phys J E 10:69
210. Knaapila M, Ikkala O, Torkkeli M, Jokela K, Serimaa R, Dolbnya IP, Bras W, ten Brinke G, Horsburgh LE, Palsson LO, Monkman AP (2002) Appl Phys Lett 81:1489
211. Valkama S, Ruotsolainen T, Kosonen H, Ruokolainen J, Torkkelo M, Serimaa R, ten Brinke G, Ikkala O (2003) Macromolecules 36:3986
212. Mäki-Ontto R, de Moel K, de Odorico W, Ruokolainen J, Stamm M, ten Brinke G, Ikkala O (2001) Adv Mater 13:117
213. Ruotsolainen T, Turku J, Heikkilä P, Ruokolainen J, Nykänen A, Laitinen T, Torkkeli M, Serimaa R, ten Brinke G, Harlin A, Ikkala O (2005) Adv Mater 17:1048
214. Mäki-Ontto R, de Moel K, Polushkin E, van Ekenstein GA, ten Brinke G, Ikkala O (2002) Adv Mater 14:357
215. Tiitu M, Volk N, Torkkeli M, Serimaa R, ten Brinke G, Ikkala O (2004) Macromolecules 37:7364
216. Kosonen H, Valkama S, Ruokolainen J, Knaapila M, Torkkeli M, Serimaa R, Mokman AP, ten Brinke G, Ikkala O (2003) Synth Met 137:881
217. Tsao SC, Chen HL (2004) Macromolecules 37:8984
218. Pollino JM, Stubbs LP, Weck M (2003) Macromolecules 36:2230
219. Stubbs LP, Weck M (2003) Chem Eur J 9:992
220. Nai KP, Pollino JM, Weck M (2006) Macromolecules 39:931
221. Binder WH, Kluger C (2004) Macromolecules 37:9321
222. Burd C, Weck M (2005) Macromolecules 38:7225
223. Elkins CL, Park T, Mckee MG, Long TE (2005) J Polym Sci A Polym Chem 43:4618
224. Shenhar R, Sanyal A, Uzun O, Nakade H, Rotello VM (2004) Macromolecules 37:4931
225. Shenhar R, Xu H, Frankamp BL, Mates TE, Sanyal A, Uzun O, Rotello VM (2005) J Am Chem Soc 127:16318
226. Percec V, Bera TK (2002) Biomacromolecules 3:272
227. Chen CP, Dai SA, Chang HL, Su WC, Wu TM, Jeng RJ (2005) Polymer 46:11849
228. DeFeyter S, DeShryver FC (2003) Chem Soc Rev 32:139
229. Samori P (2005) Chem Soc Rev 34:551
230. Gottardelli G, Masiero S, Mezzina E, Pieraccini S, Rabe JP, Samori P, Spada GP (2000) Chem Eur J 6:3242
231. Yang WS, Chai XD, Chi LF, Liu XD, Cao YW, Lu R, Jiang YS, Tang XY, Fuchs H, Li TJ (1999) Chem Eur J 5:1144
232. Samori P, Ecker C, Gössl I, de Witte PAJ, Cornelissen JJLM, Metselaar GA, Otten MBJ, Rowan AE, Nolte RJM, Rabe JP (2005) Macromolecules 35:5290
233. Cui S, Zhang W, Xu Q, Wang C, Zhang X (2003) Macromol Symp 195:109
234. Decher G (1997) Science 277:1232
235. Hammond PT (2000) Curr Opin Colloid Interf Sci 5:430
236. Zhou S, Zhang Z, Förch R, Knoll W, Schönherr H, Vancso GJ (2003) Langmuir 19:8618
237. Zhou S, Schönherr H, Vancso GJ (2005) Angew Chem Int Ed 44:956
238. Motesharei K, Myles DC (1998) J Am Chem Soc 120:7328
239. Zirbs R, Kienberger F, Hinterdorfer P, Binder WH (2005) Langmuir 21:8414
240. Binder WH, Kluger C, Straif CJ, Friedbacher G (2005) Macromolecules 38:9405
241. Boal AK, Rotello VM (2002) J Am Chem Soc 124:5019

242. Credo GM, Boal AK, Das K, Galow TH, Rotello VM, Feldheim DL, Gorman CB (2002) J Am Chem Soc 124:9036
243. Frankamp BL, Boal AK, Rotello VM (2002) J Am Chem Soc 124:15146
244. Jeoung E, Carroll JB, Rotello VM (2002) Chem Commun 1510
245. Sanyal A, Norsten TB, Uzun O, Rotello VM (2004) Langmuir 20:5958
246. Xu H, Norsten TB, Uzun O, Jeoung E, Rotello VM (2005) Chem Commun 5157
247. Viswanathan K, Ozhalici H, Elkins CL, Heisey C, Ward TC, Long TE (2006) Langmuir 22:1099

Adv Polym Sci (2007) 207: 79–112
DOI 10.1007/12_2006_110

Published online: 30 January 2007

Assembly via Hydrogen Bonds of Low Molar Mass Compounds into Supramolecular Polymers

Laurent Bouteiller

Université Pierre et Marie Curie - Paris 6,
UMR 7610 CNRS - Laboratoire de Chimie des Polymères, 4 place Jussieu,
75252 Paris Cedex 05, France
bouteil@ccr.jussieu.fr

1 **Introduction** 81

2 **Macroscopic Properties of HBSPs** 82
2.1 Rheological Properties of HBSP Solutions 82
2.1.1 Ureidopyrimidinone (UPy) 82
2.1.2 Benzene-Tricarboxamide (BTC) 83
2.1.3 Cyclohexane-Tricarboxamide (CTC) 84
2.1.4 Bis-Urea 85
2.1.5 Oligopeptides 90
2.2 Material Properties of Bulk HBSPs 91
2.2.1 Amorphous Glasses 91
2.2.2 Macroscopic Fiber Formation 91
2.2.3 Elastic Materials 92
2.3 Liquid Crystallinity 93

3 **Engineering Possibilities** 94
3.1 Improving the Strength of the Association 94
3.2 Influence of the Solvent 96
3.3 Tuning the Ring-Chain Equilibrium 97
3.4 Copolymers 97
3.5 Introducing Branches or Crosslinks 98
3.6 Responsiveness Induced by External Triggers 98
3.7 Chirality 99
3.8 Coupling Electro-Optical Properties 100
3.9 Polarity of the Chain 100
3.10 Chain Stoppers 101
3.11 Surface Grafting 101
3.12 Covalent Capture 102

4 **Molar Mass Measurement** 102
4.1 Size Exclusion Chromatography (SEC) 104
4.2 Light Scattering 104
4.3 Small Angle Neutron Scattering (SANS) 104
4.4 Viscosimetry 104
4.5 Vapor Pressure Osmometry (VPO) 105
4.6 NMR Spectroscopy 105
4.7 FTIR Spectroscopy 105

4.8 Fluorescence Spectroscopy 105
4.9 Isothermal Titration Calorimetry (ITC) 106

5 Conclusions and Outlook 106

References 106

Abstract Supramolecular polymers are linear chains of low molar mass monomers held together by reversible and highly directional non-covalent interactions. In suitable experimental conditions, they can display polymer-like rheological or mechanical properties, because of their macromolecular architecture. However, the fact that non-covalent interactions are involved means that the assembly can be reversibly broken and can be under thermodynamic equilibrium. This reversibility brings additional features compared to usual polymers, which potentially lead to new properties, such as improved processing, self-healing behavior or stimuli responsiveness. The present chapter focuses first on particular examples where macroscopic properties of HBSPs are clearly demonstrated, and then on the numerous engineering options explored so far to obtain functional materials. Finally, because the obtained properties depend strongly on the molar mass of the supramolecular polymer in the conditions of use, the last part describes the techniques available to characterize the molar mass of supramolecular polymers.

Keywords Hydrogen bond · Molar mass · Rheology · Self assembly · Supramolecular polymer

Abbreviations

AFM	Atomic force microscopy
BTC	Benzene-tricarboxamide
C*	Overlap concentration
CTC	Cyclohexane-tricarboxamide
DP_n	Number average degree of polymerization
DSC	Differential scanning calorimetry
FTIR	Fourier transform infrared spectroscopy
HBSP	Hydrogen-bonded supramolecular polymer
ITC	Isothermal titration calorimetry
M_n	Number average molar mass
NMR	Nuclear magnetic resonance
PDMS	Poly(dimethylsiloxane)
SANS	Small angle neutron scattering
SEC	Size exclusion chromatography
STM	Scanning tunneling microscopy
T_g	Glass transition temperature
UPy	Ureidopyrimidinone
VPO	Vapor pressure osmometry

1 Introduction

Supramolecular polymers are linear chains of low molar mass monomers held together by reversible and highly directional non-covalent interactions. In suitable experimental conditions, they can display polymer-like rheological or mechanical properties, because of their macromolecular architecture. However, the fact that non-covalent interactions are involved means that the assembly can be reversibly broken. This reversibility brings additional features compared to usual polymers, which can potentially lead to new properties, such as improved processing, self-healing behavior or stimuli responsiveness.

It is possible to find in the literature early examples describing the formation of hydrogen-bonded oligomers from simple monomers such as 4-(thio)pyridone [1,2] or dialkylureas [3,4]. However, the concept of supramolecular polymers was really demonstrated in 1990 by Lehn et al., who prepared a liquid crystalline supramolecular polymer by self-assembly of two complementary monomers [5]. Moreover, a decade ago, Meijer et al. described the first supramolecular polymer which formed highly viscous dilute solutions [6], thus proving that it is possible to obtain polymer-like rheological properties. Usually, the monomers can be schematized as two (or more) hydrogen bonding moieties linked through a spacer (Fig. 1). According to the topology of the hydrogen bonding groups, three main classes of hydrogen-

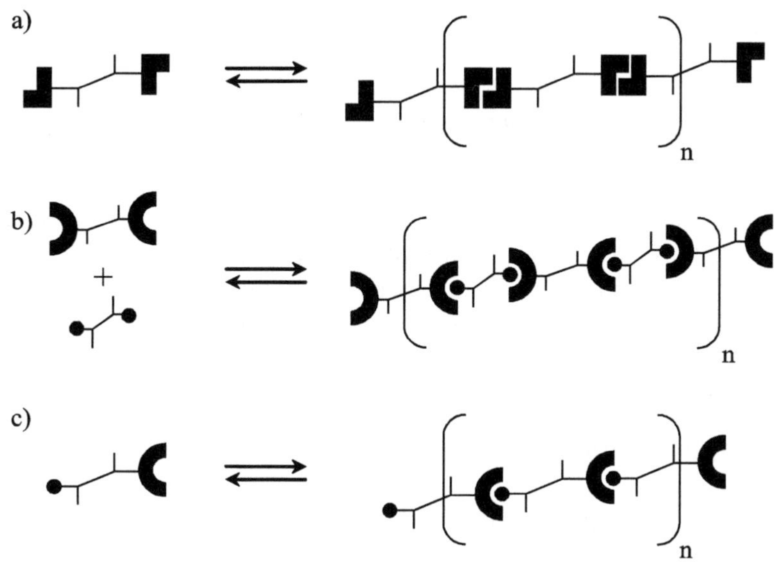

Fig. 1 Schematic representation of supramolecular polymers assembled from self-complementary AA (**a**) or AB (**c**) monomers or complementary A – A + B – B (**b**) monomers

bonded supramolecular polymers (HBSPs) can be found: self-complementary A – A or A – B monomers, or complementary A – A + B – B monomers.

In some cases, the term "supramolecular polymer" is given a wider meaning, encompassing the huge field at the crossroad between supramolecular chemistry and polymer science [7–10]. In the present chapter, we stick to the more restricted definition mentioned above: we focus on synthetic hydrogen bonded low molar mass compounds which self-assemble to form dynamic polymer-like chains. Thus, the related fields of organogelators [11, 12], nanofibers [13–15] or crystal engineering [16, 17], where crystal packing forces play a major role, are not covered. Likewise, systems where the main driving force for assembly is electrostatic, metal-ligand, hydrophobic or π-stacking interactions are not systematically included, even if some hydrogen bonding is involved. In fact, HBSPs have been reviewed in the past [18–22], but the fast development of this field justifies the present work, which concentrates on the last few years. This chapter focuses first on examples where macroscopic properties of HBSPs are clearly demonstrated, and then on the numerous engineering options explored so far. The last part describes the techniques available to characterize the molar mass of supramolecular polymers.

2 Macroscopic Properties of HBSPs

Three kinds of properties are considered here: rheological properties of solutions, properties of bulk materials and liquid crystallinity.

2.1 Rheological Properties of HBSP Solutions

Depending on the system, different rheological behaviors can be obtained. However, a common requirement is that the association must be very stable for the HBSP to have a significant molar mass in dilute conditions (Sect. 4). The following examples have been chosen because of the large amount of rheological data available, and have been grouped according to the nature of the hydrogen bonding moiety driving the association.

2.1.1 Ureidopyrimidinone (UPy)

The quadruple hydrogen bonding motif of UPy (Fig. 2a) has been designed to form very strong dimers ($K_{dim} = 2 \times 10^7$ L mol^{-1} in chloroform at 25 °C) [23–25]. Consequently, difunctional monomer **1** (Fig. 2b) forms long chains even in dilute solutions: from the value of the equilibrium constant, a degree

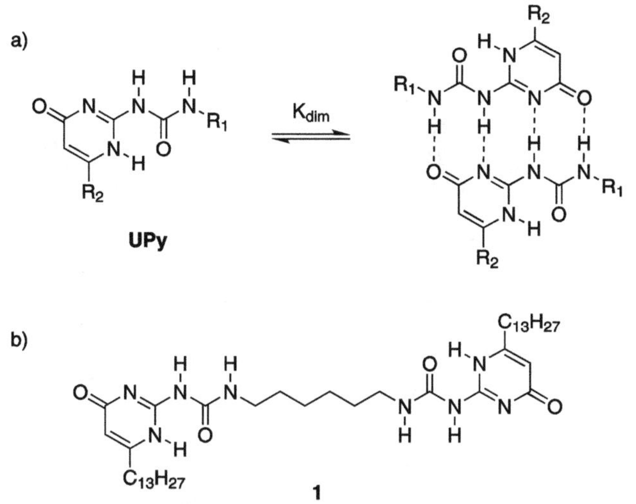

Fig. 2 Structure of UPy dimer (**a**) and monomer **1** (**b**)

of polymerization of $DP_n = 1800$ ($M_n = 1.3 \times 10^6$ g mol^{-1}) at 0.04 mol L^{-1} (30 g L^{-1}) in chloroform can be estimated. It is then not surprising that these solutions show a high viscosity ($\eta/\eta_0 = 12$ at a concentration $C =$ 0.04 mol L^{-1}) and a high concentration dependence of the viscosity ($\eta/\eta_0 \sim C^{3.7}$) [6]. The value of this exponent is in agreement with Cates's model for reversibly breakable chains above the overlap concentration [26, 27].

2.1.2 Benzene-Tricarboxamide (BTC)

Several BTC derivatives (Fig. 3a) have been shown to form viscoelastic solutions in non-polar solvents such as n-alcanes [28]. Based on the crystalline structure of a model compound [29], a supramolecular structure has been proposed. In this proposed structure, monomers are stacked onto each other due to the formation of three hydrogen bonds between the amide groups and to π-stacking between the aromatic groups (Fig. 3a). Because aromatic and amide groups tend to favor a coplanar conformation, the hydrogen bonds do not lie parallel to the column axis, but are tilted. Thus, the hydrogen-bond pattern is believed to be helical [30, 31]. However, the presence of a significant fraction of free NH groups detected by FTIR spectroscopy suggests that many defects are present in this helical hydrogen bonding pattern [32].

The consequence of this organization at the molecular level is that BTC solutions in decane are viscoelastic fluids with a nearly perfect Maxwellian behavior [33]. The reason why this BTC system is viscoelastic whereas the previous UPy based system is purely viscous (Newtonian behavior) is probably related to a more rigid backbone and/or to a slower breaking of the chains.

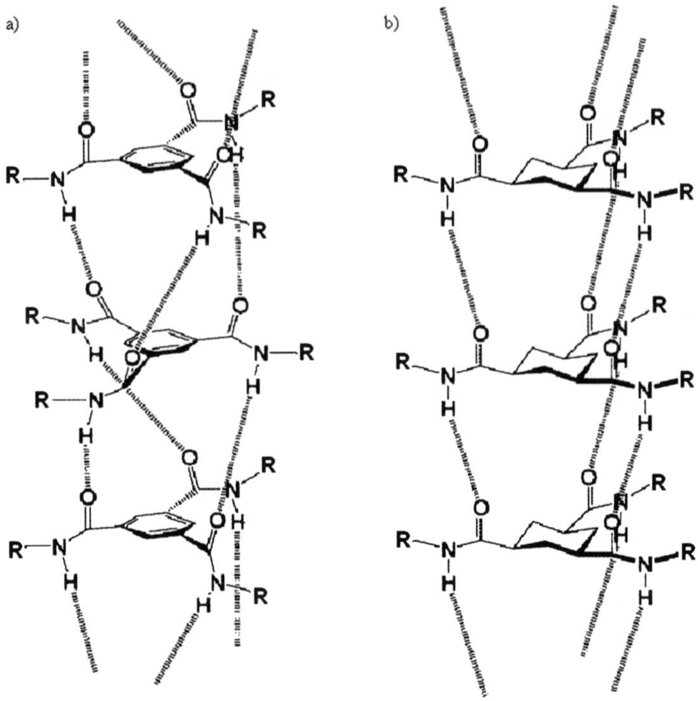

Fig. 3 Self-assembled structure of BTC (**a**) and CTC (**b**) supramolecular polymers (R = alkyl group). Adapted with permission from [34]

2.1.3 Cyclohexane-Tricarboxamide (CTC)

Some related CTC derivatives (Fig. 3b) also form viscoelastic solutions in several non-polar solvents [35], but the rheological signature is different from the case of BTC: the frequency dependence of the storage and loss moduli cannot be described by a single relaxation time [36]. Two relaxation times are necessary to adequately fit the data, so that the authors propose a model where the supramolecular polymer would present alternative sequences of rigid rod-like parts and more flexible parts. The increased rigidity of CTC compared to BTC is attributed to its particular hydrogen bonding pattern; because of the lack of π-stacking interaction and the lack of conjugation between the amides and the cyclohexane ring, the hydrogen bonds are believed to be parallel to the column axis (Fig. 3b). This hypothesis is supported by X-ray crystallography of a model compound [37]. The straight hydrogen bonding pattern of CTC may then lead to fewer defects (and thus more rigidity) than the helical pattern of BTC, because no helix reversal defects are expected.

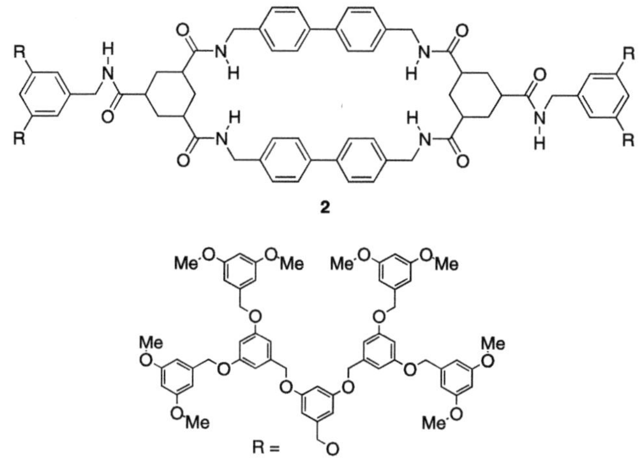

Fig. 4 Structure of monomer 2

Finally, it is worth mentioning the synthesis of monomer **2** bearing two such CTC moieties (Fig. 4) [34]. The straight hydrogen bonding pattern of each CTC moiety is compatible with the formation of six parallel rows of hydrogen bonds, so that compound **2** self-associates strongly in chloroform.

2.1.4
Bis-Urea

A variety of compounds bearing only two urea functions have been shown to form stable supramolecular architectures, because ureas can form stronger hydrogen bonds than amides. If a parallel or antiparallel orientation of the two ureas is enforced by the spacer connecting them, then long one-dimensional supramolecular assemblies can be expected. Depending on the exact nature of the spacer and the lateral substituents, it is possible to tune both the structure and the dynamic character of the assemblies. With symmetrical spacers and regular substituents, crystallization of the bis-urea is favored, so that organogelators can be obtained [11, 12, 38]. These compounds are dissolved at high temperatures in a particular solvent, but after cooling, highly anisotropic crystalline fibers are formed and entrap the solvent. The strong gels obtained are metastable and no dynamic exchange between the fibers occurs at room temperature. However, using an unsymmetrical spacer and/or branched substituents, one can try and destabilize competing crystalline structures and stabilize dynamic HBSPs. Bis-ureas **3** to **6** with a 2,4-toluene spacer (Fig. 5) indeed form dynamic supramolecular polymers in non-polar solvents [39].

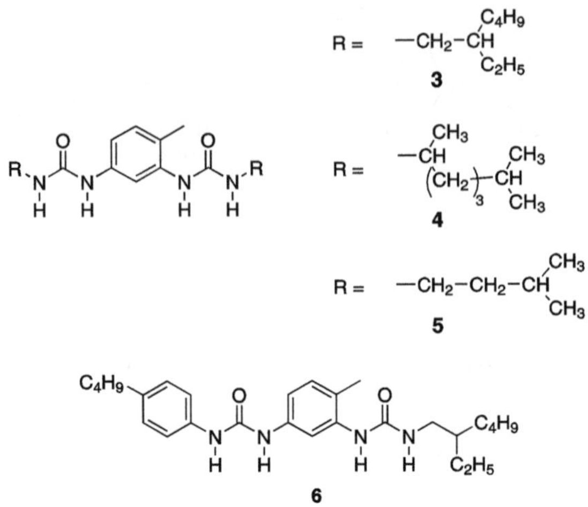

Fig. 5 Structure of bis-ureas **3**, **4**, **5** and **6**

2.1.4.1
Supramolecular Structure of Bis-Urea 3

Figure 6 shows the pseudo-phase diagram of bis-urea **3** in toluene [40]. It is likely that other (not necessarily dynamic) supramolecular structures exist at

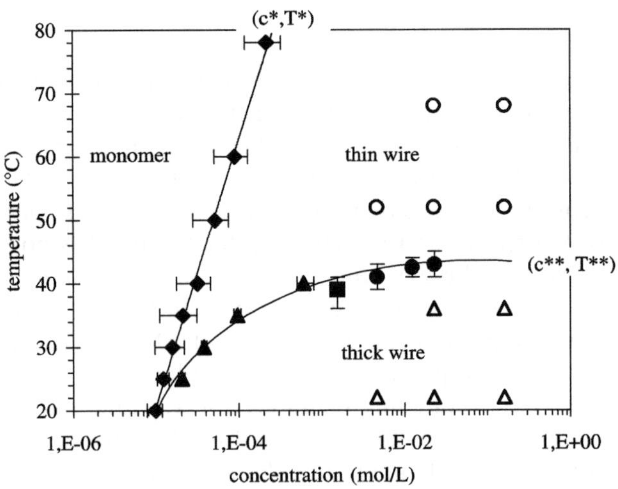

Fig. 6 Pseudo-phase diagram for supramolecular polymer **3** in toluene solutions. Transition between monomers and thin supramolecular filaments determined by calorimetry (ITC) (♦). Transition between thin filaments and thick tubes determined by ITC (▲), viscosimetry (■) and FTIR (•). SANS characterization of the thin filaments (○) and thick tubes (△). Reprinted with permission from [40]

lower temperatures or at higher concentrations, but the remarkable feature about this system is that it displays two distinct supramolecular architectures, which are stable over a wide range of concentrations and temperatures, and are in dynamic exchange with the monomer. Of course, the lines on this diagram are not true phase transitions, but limit the domains where each structure is the most abundant. For both supramolecular structures, FTIR spectroscopy can detect no free hydrogen bond. Moreover, small angle neutron scattering (SANS) shows that both structures are long and fibrillar (Fig. 7), the high temperature structure being thinner than the low temperature structure. Based on the SANS derived dimensions, on molecular simulation and on the structure of a monolayer probed by STM (Fig. 8) [41], a ladder-like supramolecular arrangement has been proposed for the high temperature, thin filament structure (Fig. 9a) [40]. Similarly, a thick tubular arrangement has been proposed for the low temperature structure (Fig. 9b,c) [42]. Such a dynamic tubular structure can be expected to be stable only if the inner cav-

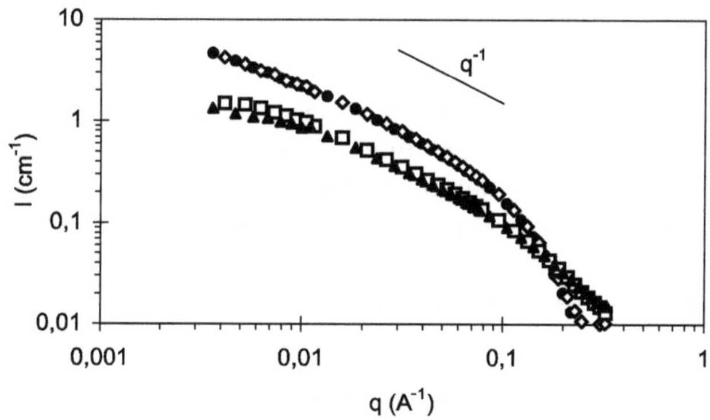

Fig. 7 SANS intensity (I) versus momentum transfer (q), for a 22.9 mM solution of supramolecular polymer **3** in d_8-toluene at several temperatures (22 °C (•); 36 °C (◊); 52 °C (□); 68 °C (▲)). Reprinted with permission from [40]

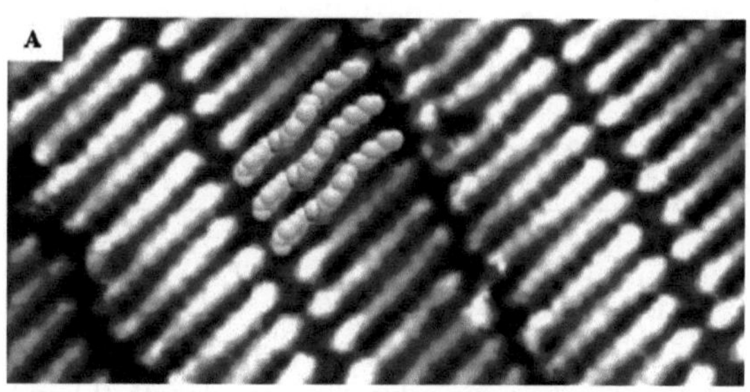

Fig. 8 High resolution STM image of a monolayer of supramolecular polymer **3** on Au(111) ($5 \times 10\,nm^2$, – 0.4 V, 1.9 nA), with insets of a space filling model **3**. Reprinted with permission from [41]

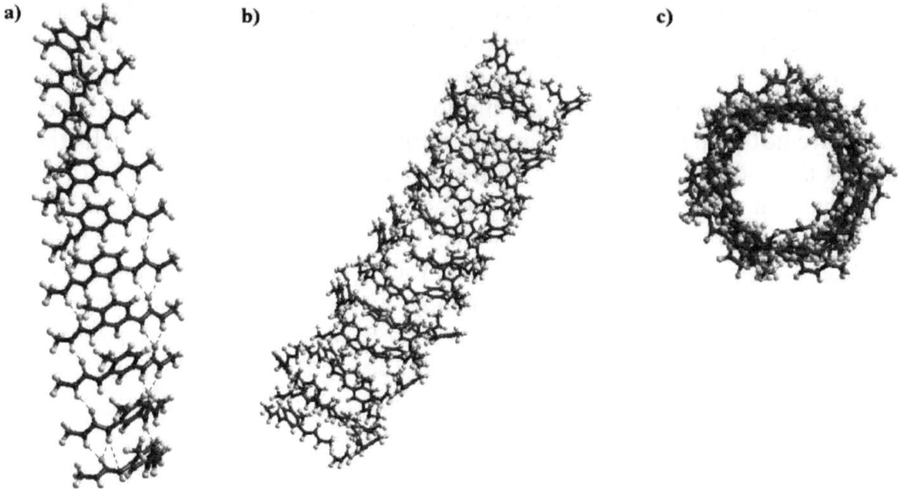

Fig. 9 Tentative supramolecular structures proposed for bis-urea **3**: thin filament (**a**) and tubular arrangements: **b** *side-view*, **c** *top-view*

ity is filled with solvent. Consequently, a very strong solvent effect is expected, with solvents of large molecular dimensions destabilizing the tubular structure. This effect was indeed demonstrated (Fig. 10) with a series of aromatic solvents of similar dielectric constants and solvating power [42]. For instance, the transition temperature between the thin and the thick structure is more than 50 °C lower in bulky 1,3,5-triisopropylbenzene than in toluene.

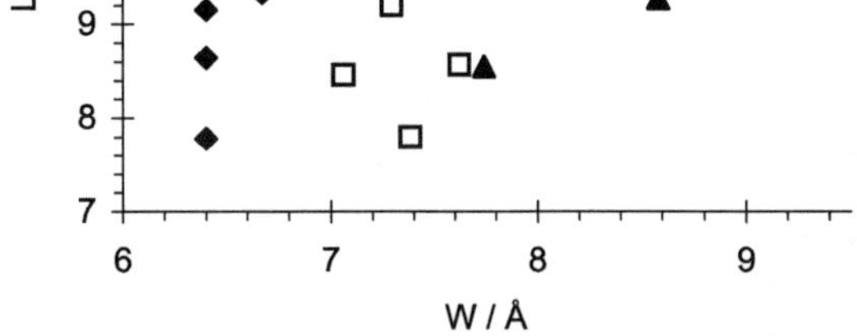

Fig. 10 Transition temperature (T^{**}) between the thin filaments and tubes for supramolecular polymer **3** solutions in aromatic solvents, versus length (L) and width (W) of the solvent molecules. The length (L), width (W) and thickness (Th) are defined as the respective dimensions of the smallest right-angled parallelepiped containing the molecule, such that $L > W > Th$. Reprinted with permission from [42]

2.1.4.2
Properties of the Thin Filament Structure

The bis-urea thin filaments can be very long in non-polar solvents such as 1,3,5-trimethylbenzene. Consequently, these solutions show a high viscosity ($\eta/\eta_0 = 8$ at a concentration $C = 0.04\ \mathrm{mol\,L^{-1}}$ and at $T = 20\ °\mathrm{C}$) and a high concentration dependence of the viscosity ($\eta/\eta_0 \sim C^{3.5}$) [43]. As in the case of UPy based supramolecular polymers, the value of this exponent is in agreement with Cates's model for reversibly breakable polymers [26, 27]. However, the solutions are not viscoelastic, even at concentrations well above the overlap concentration [43]. Consequently, the relaxation of entanglements, probably by chain scission, must be fast ($\tau < 0.01$ s).

2.1.4.3
Properties of the Tubular Structure

In contrast, the tubular structure yields strongly viscoelastic solutions in the semi-dilute regime [44] (Ducouret et al., unpublished results). Figure 11 shows a Cole–Cole plot for a dodecane solution of **3** ($C^* = 0.1\ \mathrm{g\,L^{-1}}$). Experimental data can be fitted at low frequencies with a Maxwell model, in agreement with the release of entanglements through scission and recombination, but the departure from monoexponentiality at higher frequencies is an indication that the scission-recombination of the supramolecular polymer chains may not be much faster than their reptation. Moreover, a static light scattering study on cyclohexane solutions has shown that the persistence length of the bis-urea tubes is at least 100 nm [45]. In the framework of Cates theory, the rheological characteristics of the bis-urea tubular structure can thus be explained by the presence of semiflexible filaments for which the breaking and reptation times are of the same order of magnitude [46].

In the non-linear regime, the bis-urea solutions display stress-strain curves typical of shear-banding (Ducouret et al., unpublished results).

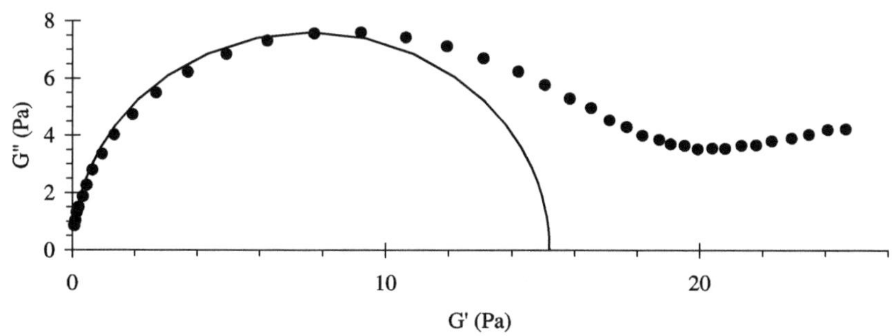

Fig. 11 Cole–cole plot for a 7.8 $\mathrm{g\,L^{-1}}$ solution of supramolecular polymer **3** in dodecane, at 25 °C

In summary, the rheological properties of these bis-urea solutions can be switched from a viscoelastic behavior (at low temperatures) to a purely viscous behavior (at high temperatures). Moreover, the transition has been shown to be fast, reversible (without hysteresis) and extremely cooperative: the conversion of tubes into thin filaments occurs within a temperature range of 5 °C only [40]. This transition can be triggered by temperature, but also by a change in the solvent composition or by a change of the monomer composition.

2.1.5 Oligopeptides

Carefully designed oligopeptides can self-assemble to form very long β-sheet tapes (Fig. 12). Of course, hydrogen bonding is not the only interaction involved, but if the β-sheets do not further crystallize into irreversible fibers, then these oligopeptides can be considered to be HBSPs, as defined in the Introduction. Boden et al. have indeed prepared several such oligopeptides which form dynamic antiparallel β-sheet tapes at very low concentrations in alcohols or in water [47,48]. At higher concentrations, the tapes dimerize into twisted ribbons due to side-chain interactions, and at still higher concentrations, the ribbons further assemble into fibrils of discrete thickness (Fig. 13) [49,50]. The different stages of the assembly can be controlled by changing the concentration or the pH.

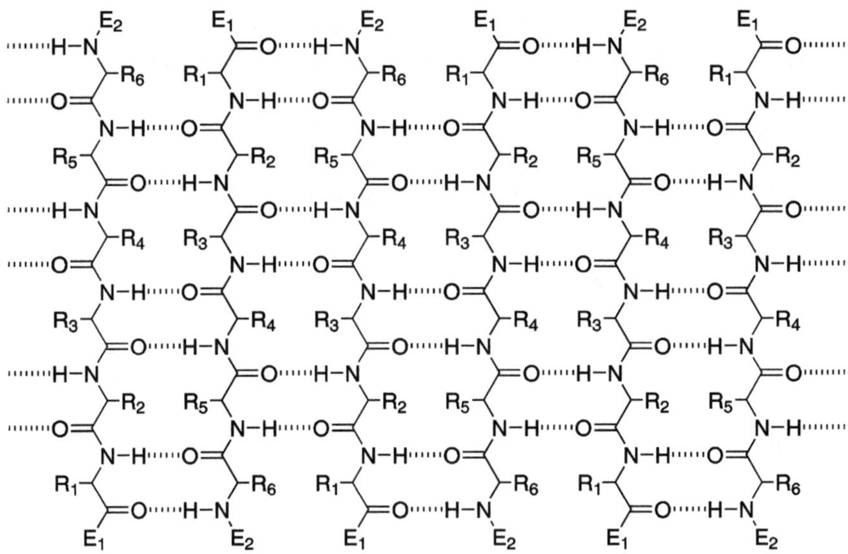

Fig. 12 Schematic representation of the self-assembly of a six-residue peptide to form a growing antiparallel β-sheet tape

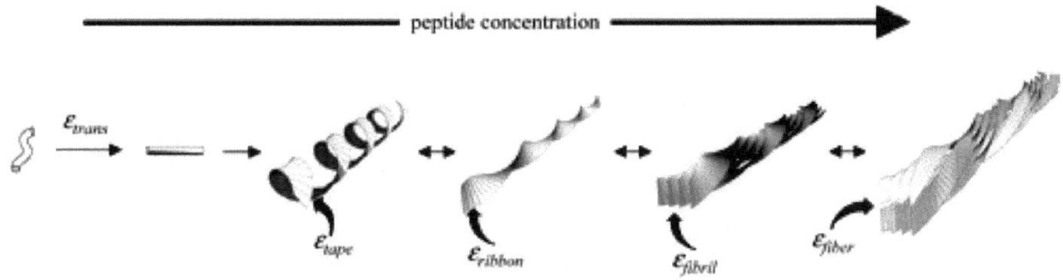

Fig. 13 Hierarchical self-assembly of β-sheet forming peptides. Reprinted with permission from [50]

Small-strain oscillatory shear experiments show that the β-sheet tapes form elastic gels over the whole frequency window (10^{-2}–10^{2} rad s^{-1}), implying that the relaxation time of the network is very long [48].

2.2 Material Properties of Bulk HBSPs

In the absence of solvent, low molar mass compounds tend to crystallize. Therefore, if polymer-like properties are desired, it is necessary to reduce at least partly their crystallization tendency, through adequate molecular design. The following examples were chosen to illustrate the range of properties currently achieved with HBSPs.

2.2.1 Amorphous Glasses

A few low molar mass compounds form molecular glasses [51], even without strong specific interaction between molecules. However, such compounds are quite rare, and the stability of their glass is usually low. A good way to improve the glass forming ability of low molar mass compounds is to introduce hydrogen bonding groups [52, 53]. For instance, mixing bisphenol-A with tetrapyridine **8** (Fig. 14) in a 2 : 1 ratio yields a stable glass with a glass transition of 31 °C, whereas the pure components are crystalline [52]. Another example is provided by the family of rigid tetrahedral compounds **9** (Fig. 15). If the substituent R is a butyl group, the compound crystallizes, but if R is a longer alkyl group, amorphous solids are obtained, probably because the steric hindrance of the alkyl group introduces sufficient disorder [53]. In this case, high T_gs are obtained (T_g = 135 °C for R = hexyl), due to the rigidity of the compounds. Above the glass transition, high viscosity Newtonian fluids are obtained.

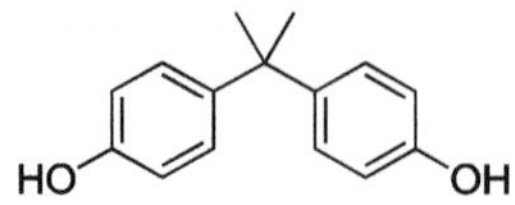

7

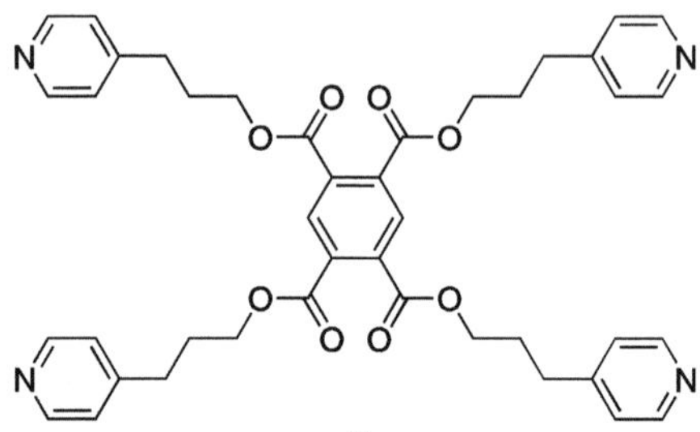

8

Fig. 14 Structure of bisphenol-A **7** and tetrapyridine **8**

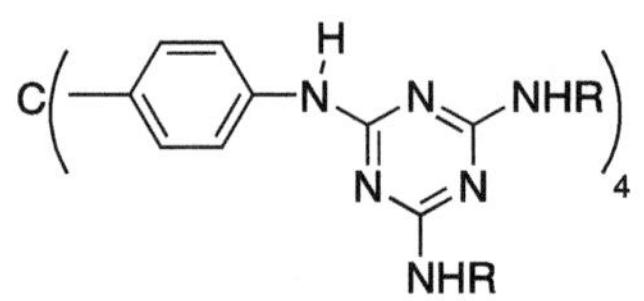

9

Fig. 15 Structure of tetrahedral monomer **9**

2.2.2 Macroscopic Fiber Formation

When hydrogen bonds are established along a preferred direction, it is possible to extrude reasonably strong fibers from the melt [54–56] or from a concentrated solution [57–59]. For example, utilizing the CTC backbone (Sect. 2.1.3) substituted by branched alkylsilyl sidechains, Araki et al. succeeded in spinning fibers with tensile strengths in the 1 MPa range [56].

2.2.3 Elastic Materials

More interestingly, functionalizing oligomers with strongly dimerizing units like UPy (Sect. 2.1.1) yields materials with elastic properties similar to high molar mass polymers. This approach is successful even in the case of a very short siloxane oligomer: compound **10** (Fig. 16) displays a narrow rubbery plateau between its glass transition temperature ($T_g = 25\,°C$) and 70 °C, whereas non-hydrogen bonded reference compound **10-Bn** is a crystalline

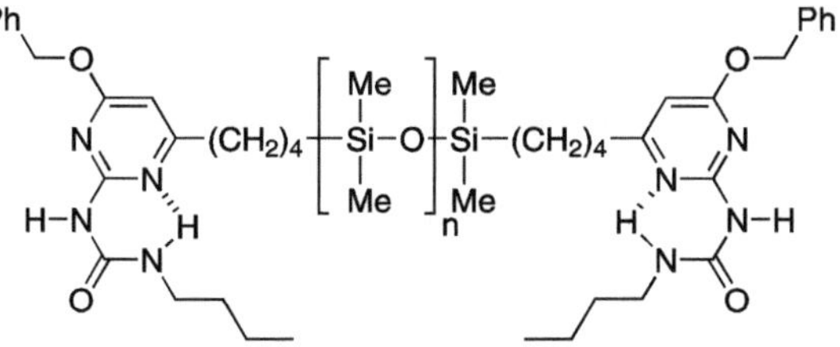

10-Bn : n=2
11-Bn : n=100

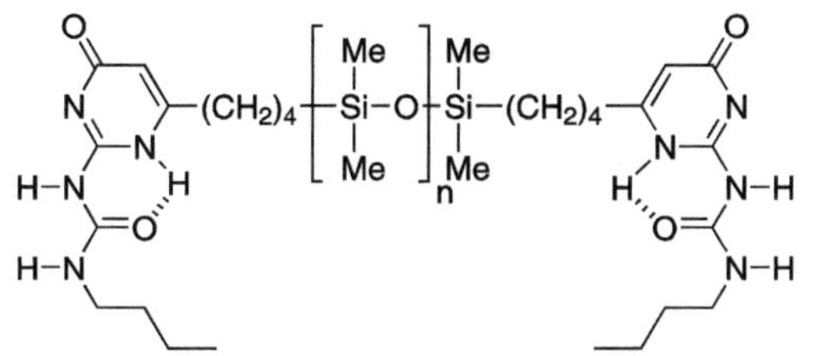

10 : n=2
11 : n=100

Fig. 16 Structure of monomers **10-Bn**, **10**, **11-Bn** and **11**

solid below and a low viscosity liquid above its melting point [60]. With a longer spacer, PDMS **11** shows even better properties. For instance, at low frequencies the complex viscosity of **11** is 2000 times the viscosity of **11-Bn**. Based on DSC and solid-state NMR experiments, it seems that this effect is only due to dimerization of chain-ends, and not to any microcrystalline domains [60]. This approach has been extended to the case of short telechelic poly(ethylene/butylene), polyether, polyester and polycarbonate oligomers [61]. The strong temperature dependence of the viscosity clearly gives a processing advantage to these materials. Moreover, the good thermal stability of the UPy group is to be noted [62].

2.3 Liquid Crystallinity

The use of hydrogen bonds in liquid crystalline materials at large is a very active area [63, 64]. In the case of main chain thermotropic liquid crystalline HBSPs, two main approaches have been explored. The first approach is to mix two complementary monomers, usually of the AA + BB type (Fig. 1). The complementary hydrogen bonding units utilized are mainly modified nucleobases [5, 65, 66] and aromatic acid/pyridine couples [54, 67–74]. In the most successful cases, the pure components are not mesomorphous, but the mixtures are [65, 69–71]. At any rate, the stability range of the mesophase is usually increased. The second approach is to use a self-complementary monomer which self-assembles into columns stabilized by amide [75–83] or urea [84] bonds along the column axis. Sierra et al. recently proposed a system, which can be considered to be intermediate between the two approaches [85].

Lyotropic HBSPs have also been described [57].

3 Engineering Possibilities

HBSPs self-assemble because the monomers contain specific and directional complementary associating groups. In fact, the structure of a monomer can be understood to consist of two independent parts: the associating groups, which can be engineered to optimize the self-assembly process, and the remaining of the molecule (Fig. 1). The latter can be altered nearly at will without compromising the self-assembly, as long as no interfering hydrogen bonding groups are introduced. Consequently, it is possible to tune the properties and to add functionality through chemical design. The following examples illustrate this point.

3.1 Improving the Strength of the Association

The quantitative treatment of the growth of supramolecular chains is discussed in Sect. 4, but for the present discussion, a qualitative approach is sufficient. Usual associating groups such as acid, amide or urea functions can only form one or two hydrogen bonds and are, therefore, quite weak. It seems intuitive that the strength of the association increases with the number of hydrogen bonds involved in the assembly. This is indeed a major parameter, but not the only one. Based on studies of monofunctional compounds, it has been established that the following parameters cannot be neglected: secondary electrostatic interactions, preorganization of the recognition unit and the presence of competing tautomers [25, 86]. Secondary electrostatic interactions (Fig. 17) arise from the repulsion or attractions between partial charges

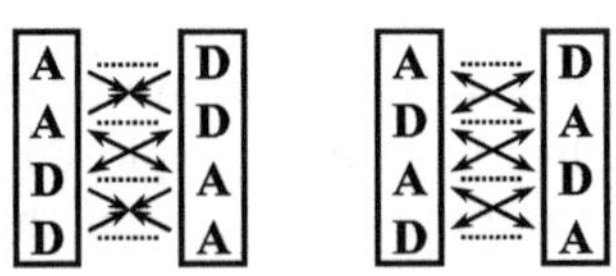

$3.6\ 10^6$ L mol^{-1} $3.1\ 10^2$ L mol^{-1}

Fig. 17 Self-complementary dimers formed by linear arrays of four hydrogen bonding sites (A: acceptor, D: donor), and their stability constants in $CDCl_3$ as predicted in reference [88]. Attractive and repulsive secondary interactions are indicated by *arrows*

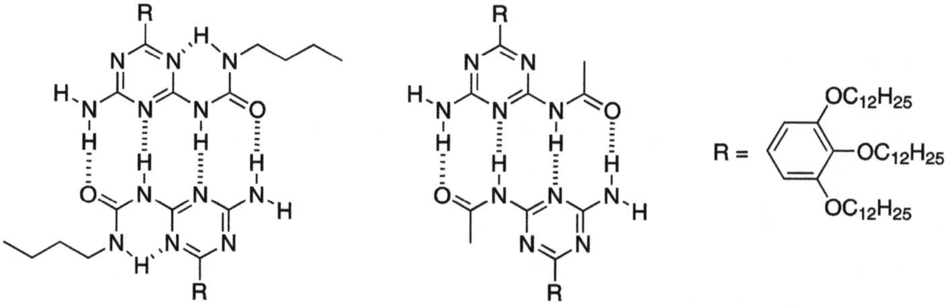

12 $K_{dim} = 2\ 10^4$ L mol^{-1}

13 $K_{dim} = 530$ L mol^{-1}

Fig. 18 Structure of self-complementary compounds **12** and **13**, and their dimerization constants in $CDCl_3$ [25]

localized on heteroatoms and hydrogens [87, 88]. An example of the effect of preorganization is shown on Fig. 18, where the ureidotriazine **12** dimerizes more strongly than the amidotriazine **13**, because the intramolecular hydrogen bond in **12** stabilizes the conformation suitable for association. However, Fig. 19 shows that the gain in preorganization may be offset by the influence of tautomerism, because compounds **14** and **15** display similar dimerization efficiencies [89]. The self-association of UPy derivative **14** is enhanced by the presence of the intramolecular hydrogen bond, but is weakened by the existence of three different tautomeric forms, one of which cannot dimerize [25]. On the other hand, cytosine derivative **15** is conformationally flexible, but does not undergo tautomeric changes. In fact, it is difficult to rank the influence of these different effects, because any structural change also modifies the overall charge distribution of the recognition unit. Finally, obvious steric bulk effects have also been demonstrated in the case of simple dialkylureas [90].

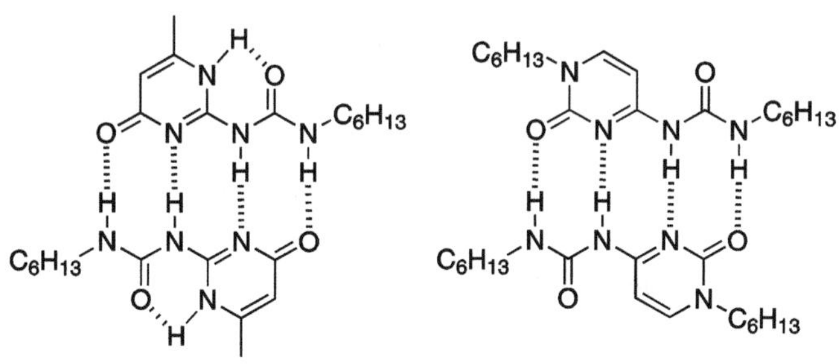

14 **15**

Fig. 19 Structure of self-complementary compounds **14** and **15**

3.2
Influence of the Solvent

Because of its electrostatic component, the hydrogen bond is affected by the solvent polarity: the lower the polarity of the solvent, the stronger the association [91]. Quantitatively, this effect can be quite significant for HBSPs: for instance the self-association constant of bis-urea **3** is two orders of magnitude larger in toluene than in chloroform [92]. This translates into a tenfold effect on the degree of polymerization.

In addition to this classical polarity effect, the solvent can have a more subtle influence. In the case of monomer **16** (Fig. 20) hydrogen bonding in chloroform leads to the formation of a usual flexible supramolecular polymer. However, in dodecane the dimerization of the ureidotriazine is reinforced by a solvophobic stacking of the aromatic parts, which yields a columnar architecture [93]. A similar solvophobic effect has been demonstrated with a UPy-based monomer [94]. Another possible "side effect" of the nature of the solvent is the occurrence of specific host–guest interactions between the HBSP and the solvent [42, 95, 96].

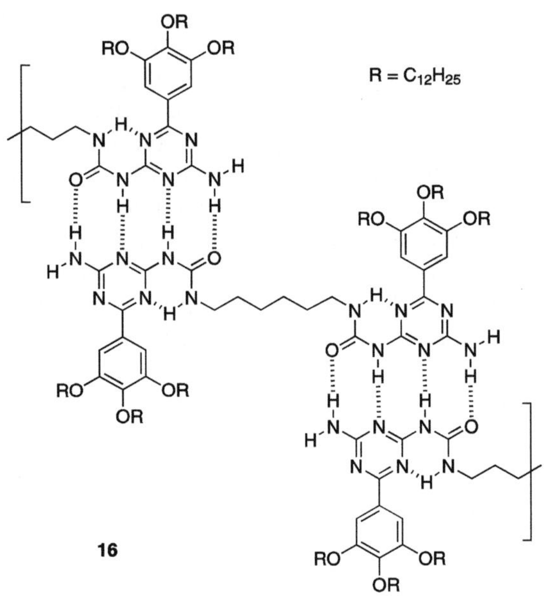

16

Fig. 20 Structure of monomer **16**

The case of water is apart due to its biological and environmental relevance. However, designing a HBSP in water is a challenge, because of the polarity and hydrogen bonding ability of water. Nevertheless, several systems based on ureas [97], peptides [47–50, 98], heteroaromatic compounds [93,

99–102], or oligonucleotides [103–106] have been successfully developed by strengthening the hydrogen bond interaction with hydrophobic, π-stacking and/or electrostatic interactions.

3.3 Tuning the Ring-Chain Equilibrium

Ring-chain equilibrium in flexible supramolecular polymers has been extensively studied both experimentally [59, 107–112] and theoretically [113, 114]. For strongly associating systems, a threshold concentration is expected below which only cyclic species are present, and above which the amount of cyclic species remains constant. It has been shown that the value of this threshold concentration, and thus the amount of cyclics, is a strong function of the length [107] and the conformation [108, 109, 113] of the spacer connecting the associating groups. In particular, for very preorganized monomers, it is possible to obtain high yields of cyclic dimers [115, 116] or tetramers [117, 118]. The physicochemical description of this situation has also been reported [119, 120]. In the case of supramolecular polymers formed by the assembly of two complementary monomers, the cyclic content is minimized if the lengths of the two monomers are mismatched [112]. Generally, a careful choice of the connecting spacer thus makes it possible to design supramolecular polymers with a given cyclisation tendency. This in turn influences the properties, mainly because of the lower molar mass of cyclics compared to linear chains.

3.4 Copolymers

In macromolecular science, copolymers are ubiquitous, because their properties can be adjusted by a change in monomer composition. In the case of supramolecular polymers, the preparation of copolymers is particularly straightforward since it requires a simple mixing of different monomers. However, in the vast majority of cases, only statistical copolymers have been reported [30, 31, 40, 71, 72]. To design more elaborate copolymers, some additional information has to be programmed in the structure of the monomers. This has been achieved for alternating HBSPs, where the alternating tendency comes from an improved complementarity [121] or from the non-homopolymerizability of a monomer, either for steric reasons [56] or because of lack of self-complementarity [122]. However, up to now, copolymers with neither statistical nor alternating sequences have only been reported for oligonucleotide-directed assembly [105, 106]. In this case, the availability of a vast number of orthogonal oligonucleotide recognition groups makes it possible to design copolymers with virtually any sequence.

3.5 Introducing Branches or Crosslinks

The use of monomers bearing more than two associating groups is a straightforward way to introduce a controlled amount of branches or crosslinks in a supramolecular polymer structure [6, 58, 121, 123–127]. The improvement of the mechanical properties can be spectacular. For instance, trifunctional monomer **17** (Fig. 21) forms highly viscous solutions in chloroform, and is a viscoelastic material in the absence of solvent [124]. The reversibly cross-linked network displays a higher plateau modulus than a comparable covalently cross-linked model. This is explained by the fact that the reversibly cross-linked network can reach the thermodynamically most stable conformation, whereas the covalent model, which has been cross-linked in solution and then dried, is kinetically trapped.

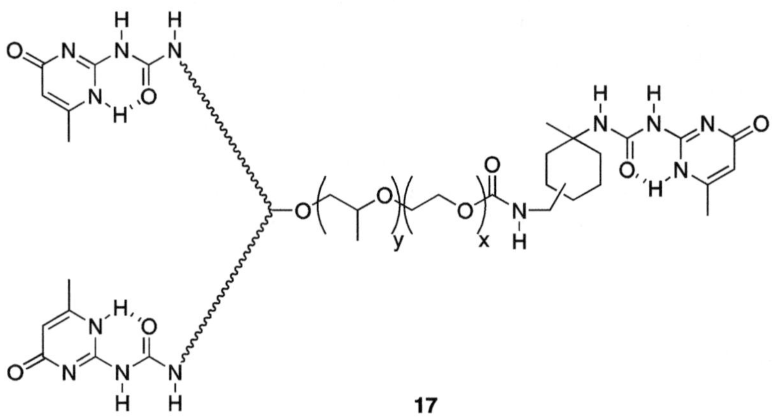

17

Fig. 21 Structure of trifunctional monomer **17**

3.6 Responsiveness Induced by External Triggers

HBSPs are intrinsically more responsive than polymers, due to the temperature and concentration dependence of their molar mass. Moreover, supramolecular polymers with added responsiveness can result from various functional elements introduced within the monomer structure. For instance, light controlled supramolecular polymers have been obtained by incorporating a photoresponsive chromophore between the two self-associating groups of the monomer. Thus, light can trigger a conformational switch [128–131] or the formation of a reversible bond within the monomer [132], which then leads to a change of the length of the chains. Alternatively, it is possible to use a chain stopper (Sect. 3.10) bearing a photocleavable protecting group [133]. The deprotection improves the efficiency of the chain stopper, so that the viscosity of the solution decreases upon irradiation.

Another approach is to introduce a suitable chemical function in between the two hydrogen bonding groups of the monomer. For example, this can be an organometallic complex cleaved by addition of a suitable ligand [134, 135]; or a reversible covalent bond [136]; or an ionic interaction controlled by the presence of CO_2 [137–139] or pH [140].

3.7 Chirality

Monomers decorated with chiral side-chains can induce a transfer of chirality at the supramolecular level. Indeed, several examples show that disk-like molecules, designed to pile up into long reversible columns, form in fact helical columns, driven by the favorable packing of the chiral lateral substituents [30, 32, 93, 94, 99]. This chiral packing effect is strong enough to induce a chiral amplification: a low amount of chiral monomer mixed with a non-chiral monomer can still drive the formation of helical columns.

With a different design, where chirality is directly built in the hydrogen bond pattern (Fig. 22), Aida et al. have demonstrated that it is possible to enforce a homochiral supramolecular polymerization [141]. In this case, a mixture of L and D monomers exclusively forms supramolecular chains of polyL and polyD homopolymers instead of copolymers.

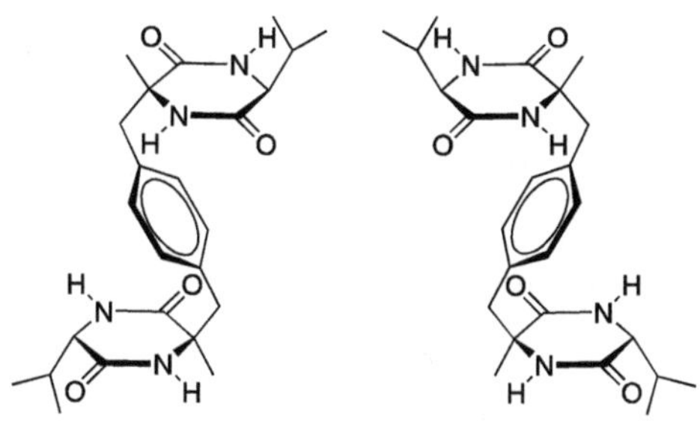

18L **18D**

Fig. 22 Structure of enantiomeric monomers **18L** and **18D**

In the case of oligopeptides which form dynamic antiparallel β-sheet tapes (Sect. 2.1.5), chirality of the amino-acids is directly responsible for the twisting of the tapes [142, 143]. Moreover, the lateral aggregation of the tapes can lead to the formation of fibrils of discrete thickness, and is in fact controlled by the twisting of the tapes [144–146].

3.8
Coupling Electro-Optical Properties

Functional supramolecular polymers have been designed by introducing particular chromophores, such as quinacridone [147], perylenes [148–150], oligophenylenevinylenes [151, 152], porphyrins [117, 153–155] or merocyanine dyes [156], within the monomer structure. The self-assembly process, by altering the distance between the chromophores, may be coupled to a particular electro-optical property (fluorescence, energy transfer). This could be of interest in the fields of light harvesting and long-range vectorial transport of excitation energy.

3.9
Polarity of the Chain

By analogy with the pointed and barbed ends of actin filaments [157], it is worthwhile designing HBSPs with two different extremities. This feature is particularly useful in the context of surface grafting of supramolecular polymers (Sect. 3.11). Dialkylureas (Fig. 23) are very simple monomers where this breaking of symmetry directly results from the structure of the monomer: one extremity exposes a free carbonyl group to the solvent, while the other extremity presents hydrogen giving groups [4, 90, 158]. More complex bow-shaped monomer **19** (Fig. 24) has also been shown to form directional assemblies [159]. In strongly associated systems like BTC [160] and CTC [161] (Sects. 2.1.2 and 2.1.3), the symmetry breaking along the chain is responsible

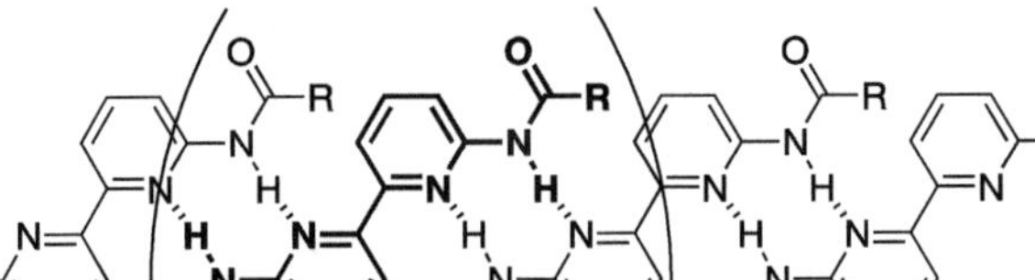

Fig. 23 Hydrogen bonding pattern of dialkyl ureas

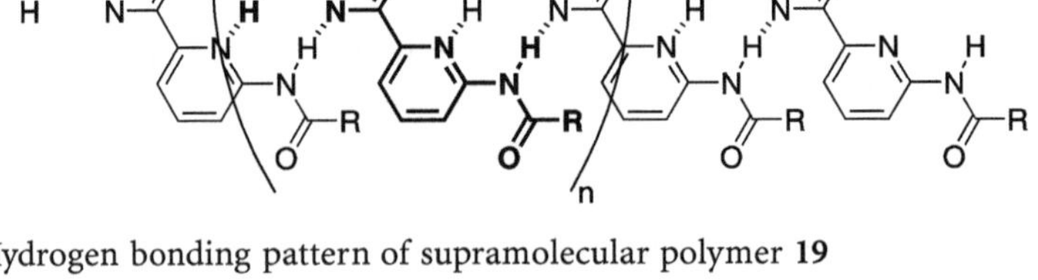

Fig. 24 Hydrogen bonding pattern of supramolecular polymer **19**

for the build up of large macrodipoles that may be useful for electrooptical or electromechanical devices.

3.10
Chain Stoppers

The fact that supramolecular polymers are assembled by highly directional interactions allows one to design chain stoppers, that is, molecules able to interact specifically with the chain ends of the filaments. For instance, a molecule bearing a single A function is expected to be a chain stopper for an A – A supramolecular polymer. Chain stoppers introduced in varying amounts have often been used simply to reduce the length of the supramolecular chains [6, 96, 104, 123, 124, 133, 154, 162, 163]. Indeed, addition of even low amounts of a chain stopper to a supramolecular polymer solution can reduce its viscosity very significantly. More interestingly, chain stoppers can be used to block the concentration dependence of the length of the supramolecular polymer, over a useful concentration range [46, 164, 165]. Indeed, if the monomer concentration and the equilibrium constant of the chain stopper are large enough, nearly all chain ends are occupied by a chain stopper, which means that the length of the filament is inversely proportional to the chain stopper fraction, and independent of the monomer concentration. This feature can be exploited to derive the molar mass and radius of gyration of the stopped supramolecular chains [164, 165]. Moreover, it means that it becomes possible to independently vary the length and the concentration of the chains, so that scaling exponents for the chain length and concentration dependence of rheological properties can be obtained, and compared to theoretical values, in order to derive some information on chain flexibility and dynamics [46].

Finally, chain stoppers can also be exploited to decorate the chain-ends with particular functional groups or labels [166].

3.11
Surface Grafting

Arguably one of the most interesting developments of the chain-stopper concept is the use as a surface anchor. Indeed, covalent grafting of a chain-stopper on a surface should yield a supramolecular polymer brush, if the surface is immersed in a supramolecular polymer solution. Such brushes have been realized experimentally with UPy [167] or oligonucleotide [168, 169] based monomers. The properties of the brushes (thickness, adhesion) have been studied by AFM, with chain-stopper grafted tips. The conclusion of these preliminary studies is that specific molecular recognition mediates direct bridging and thus adhesion between the surface and the tip. Moreover, the average length of the grafted chains seems to be shorter than the chains in the surrounding solution.

Finally, the case of a polar A – B type supramolecular polymer chain with two complementary but different chain-ends is worth considering. If such a system is brought into contact of a surface grafted with an anchoring group bearing only A functions, then a theoretical model shows that the supramolecular brushes formed should exert repulsive forces between approaching surfaces [170, 171].

3.12 Covalent Capture

Covalent capture of supramolecular assemblies can be tricky, because the energy involved in the covalent bond formation is large compared to the stabilizing energy of the self-assembling process [172–174]. Consequently, the covalent reaction should be very carefully designed, to avoid disrupting the supramolecular structure. Moreover, in the case of supramolecular polymers, inter-chain cross-linking can be a problem. Up to now, two different approaches have been successful. Meijer et al. have polymerized columnar stacks of BTC derivatives bearing a photopolymerizable sorbate group [175, 176], and Craig et al. have captured oligonucleotide based supramolecular polymers by ligation with a DNA ligase [103]. Sol-gel chemistry has also been used, but in this case, the structure prior to covalent capture has not been characterized, so that it is not known if the final structure was present before the covalent reaction [177–179].

4 Molar Mass Measurement

The average molar mass of a supramolecular polymer is a useful information, because polymer-like properties can only be expected if long chains are really formed. Unfortunately, the dynamic nature of supramolecular polymers makes their characterization less straightforward than for usual polymers: the molar mass may change during measurement because of the measuring conditions (for instance in the case of chromatography or mass spectroscopy). Moreover, the knowledge of the molar mass in some particular condition is not really sufficient, because the molar mass changes with the experimental conditions, such as the solvent, the concentration or the temperature. Consequently, it is recommended to characterize the evolution of molar mass in a range of experimental conditions.

To do so, it is necessary to consider all the possible self-assembled species present in the medium. A convenient way to handle such a complex system is to use a theoretical model involving equilibrium constants between the different species. Determination of the equilibrium constants then makes it possible to compute the average molar mass at any concentration. For ex-

$$A_1 + A_1 \overset{K_2}{\rightleftharpoons} A_2 \qquad [A_2] = K_2\,[A_1]\,[A_1]$$

$$A_2 + A_1 \overset{K_3}{\rightleftharpoons} A_3 \qquad [A_3] = K_3\,[A_2]\,[A_1]$$

$$\ldots \qquad \ldots$$

$$A_{n-1} + A_1 \overset{K_n}{\rightleftharpoons} A_n \qquad [A_n] = K_n\,[A_{n-1}]\,[A_1]$$

$$\ldots \qquad \ldots$$

Fig. 25 Theoretical association model for a self-complementary monomer A_1

ample, in the case of a self-complementary monomer A_1, which is supposed to self-assemble into linear chains (A_n) of degree of polymerization n, the relevant theoretical model is described on Fig. 25 [180]. If necessary, the model can be extended to take into account the formation of cyclics [107, 114, 118].

Although addition to a growing chain is written as sequential, association may also occur by random association. For example, in addition to adding a monomer to a pentamer, a hexamer may form from two trimers or from a dimer and a tetramer. Since we are interested in the thermodynamic equilibrium state and not in the kinetically most favored pathway or in the mechanism of aggregate formation, the selection of expressions does not affect the final result, as long as all the possible species are included. As described in Fig. 25, this model involves an infinity of unknown parameters (K_n). Simplifying assumptions are thus required. The simplest model is the so called isodesmic model, which assumes no chain-length dependence of the equilibrium constants ($K_n = K$, for $n > 1$). In this case, the number average degree of polymerization can be very simply expressed from the equilibrium constant and the total monomer concentration C_0 [181].

$$DP_n = \frac{1 + \sqrt{1 + 4KC_0}}{2} \approx \sqrt{KC_0}\,.$$

A direct information derived from this analysis is that very large equilibrium constants are required if long supramolecular chains are sought. For instance, at a concentration of $C_0 = 10^{-2}\ \mathrm{mol\,L^{-1}}$, a degree of polymerization of 100 is only possible if the equilibrium constant is $K = 10^6\ \mathrm{L\,mol^{-1}}$.

The isodesmic model has been very successfully used in many cases [6, 107, 123, 148, 149]. However, steric or electronic effects may be responsible for significant departures from the isodesmic model. In particular, anticooperative [182] or cooperative [128, 183–185] systems have been described. The case of cooperative systems, where the formation of long chains is favored over dimerization, is most interesting because the concentration dependence of the molar mass is stronger, so that the system is potentially more responsive than an isodesmic system [185].

Let us now focus on the different experimental techniques available to characterize the molar mass of supramolecular polymers.

4.1 Size Exclusion Chromatography (SEC)

The separation of chains according to their hydrodynamic volume inside SEC columns is accompanied by their dilution. Consequently, the molar mass distribution is shifted toward lower values during the measurement. Usually, the dissociation of hydrogen-bonded chains is fast compared to the elution time, so that only qualitative information can be derived [115, 121, 141]. However, if the dissociation is sufficiently slow, it is possible to measure reliable molar mass values [186, 187].

4.2 Light Scattering

The weight average molar mass of a polymer can be deduced from static light scattering experiments, through the classical Zimm treatment which involves extrapolation to zero concentration and zero angle. In the case of supramolecular polymers, the extrapolation to zero concentration is problematic because of the dynamic character of the chains. Two approaches have been proposed to circumvent this problem. The first consists in measuring the scattered light intensity at a sufficiently low concentration and to neglect the effect of the second virial coefficient [188]. Alternatively, it is possible to evaluate the second virial coefficient by using solutions containing controlled amounts of chain stoppers [164].

4.3 Small Angle Neutron Scattering (SANS)

SANS is a useful technique to derive some information on the structure and the shape of supramolecular polymers [189]. In particular, the length of rigid-rod supramolecular polymers can be determined if it is below about 50 nm [93, 166, 190]. From the length of the rods, the molar mass can easily be deduced.

4.4 Viscosimetry

Molar mass of polymers can be deduced from intrinsic viscosity measurement through a Mark–Houwink calibration curve. This approach can be applied to supramolecular polymers if Mark–Houwink and Huggins parameters are known [191]. However, finding a suitable covalent model to estimate these coefficients is a difficult task.

4.5 Vapor Pressure Osmometry (VPO)

The measure of the osmotic coefficient by VPO is increasingly used to deduce the molar mass of supramolecular polymers [165, 192–195]. However, care must be taken to work at concentrations as low as possible to minimize other contributions than self-association to non-ideality [196]. If not, the values obtained may only be considered as orders of magnitudes.

4.6 NMR Spectroscopy

Solution NMR spectroscopy is the most widely used technique to determine the equilibrium constant of supramolecular polymers [23, 24, 59, 123, 148, 149, 158, 197], either directly or with the help of a monofunctional model compound. In the usual case of a fast exchange between free and hydrogen bonded extremities, the equilibrium constants of a particular association model are derived from the analysis of the evolution of the chemical shift versus concentration (for self-complementary functions) or versus stoichiometry (for complementary functions). Equilibrium constants in the range below $10^6\ M^{-1}$ are reliably accessible.

Estimation of the molar mass of a bulk supramolecular polymer by extrapolation of values determined in solution is always questionable. Therefore, a method based on transverse relaxation measurements of bulk samples was proposed [198]. It takes advantage of the fact that the strength of residual dipolar interactions depends on the molar mass in entangled polymer melts.

4.7 FTIR Spectroscopy

Hydrogen bonding of associative groups is often characterized by a measurable shift of an absorption band. The measure of the intensity of these bands affords the equilibrium constant [90, 107, 185, 199]. However, the use of FTIR spectroscopy is less versatile than NMR spectroscopy because solvent absorption often limits the dilution range accessible.

4.8 Fluorescence Spectroscopy

In the case of fluorescent monomers, monitoring the change of fluorescence upon association directly yields the equilibrium constant [149, 182]. Alternatively, it is possible to label the monomer with a suitable excimer forming chromophore [24] or with a pair of chromophores for fluorescence resonance energy transfer studies [200]. The clear advantage of fluorescence spectroscopy is its high sensitivity enabling the measurement of equilibrium constants as high as 10^8 L mol^{-1}.

4.9 Isothermal Titration Calorimetry (ITC)

ITC has been used to determine the equilibrium constant of supramolecular polymers formed either from complementary monomers [201] or from self-complementary monomers [92, 164]. In the former case, the A – A monomer solution is injected into the B – B monomer solution. The exchanged heat measured is proportional to the number of hydrogen bonds formed and thus directly related to the equilibrium constant. In the latter case, the self-complementary monomer is simply diluted into pure solvent. This time, the exchanged heat measured is proportional to the number of hydrogen bonds broken and related to the self-equilibrium constant. ITC is a very powerful technique because equilibrium constants as high as 10^9 L mol^{-1} are accessible [202]. Moreover, a single experiment yields the equilibrium constant together with the molar enthalpy of association ($\Delta H_{\mathrm{assoc}}$).

5 Conclusions and Outlook

HBSPs, thus, constitute a very versatile family of compounds with interesting properties, whether in the presence or in the absence of solvents. Because of the reversibility of the interactions involved, HBSPs are under thermodynamic equilibrium. Consequently, their properties can be adjusted beforehand by a careful structural design, but also in-situ by external stimuli. The obtained properties depend strongly on the molar mass of the supramolecular polymer in the conditions of use (solvent, concentration, temperature). Fortunately, several techniques to measure their molar masses are now available.

Among the many interesting current developments in this field, the properties of supramolecular polymers at interfaces have been relatively little explored so far, but seem especially promising. This includes the preparation of supramolecular polymer brushes [167–171], of monolayers for surface nanopatterning [41] or the study of the influence of supramolecular polymers on colloidal stability [163].

References

1. Beak DC, Covington JB, Smith SG, White JM, Zeigler JM (1980) J Org Chem 45:1354
2. Etter MC, MacDonald JC, Wanke RA (1992) J Phys Org Chem 5:191
3. Kozlova TV, Zharkov VV (1981) Zh Prikl Spektrosk 35:303

4. Jadzyn J, Stockhausen M, Zywucki B (1987) J Phys Chem 91:754
5. Fouquey C, Lehn J-M, Levelut A-M (1990) Adv Mater 2:254
6. Sijbesma RP, Beijer FH, Brunsveld L, Folmer BJB, Hirschberg JHKK, Lange RFM, Lowe JKL, Meijer EW (1997) Science 278:1601
7. Ciferri A (2002) Macromol Rapid Commun 23:511
8. Ciferri A (2003) J Macromol Sci C43:271
9. Ciferri A (2005) Supramolecular Polymers. Marcel Dekker, New York
10. Binder W (2005) Monatshefte Chem 136:1
11. Terech P, Weiss RG (1997) Chem Rev 97:3133
12. Terech P, Weiss RG (2005) Molecular Gels: Materials with self-assembled fibrillar networks. Kluwer, Dordrecht
13. Hartgerink JD, Zubarev ER, Stupp SI (2001) Curr Opin Solid State Mat Sci 5:355
14. Shimizu T (2003) Polym J 35:1
15. Pasini D, Kraft A (2004) Curr Opin Solid State Mat Sci 8:157
16. Desiraju GR (2002) Acc Chem Res 35:565
17. Wuest JD (2005) Chem Commun, p 5830
18. Zimmerman N, Moore JS, Zimmerman SC (1998) Chem Ind 604
19. Brunsveld L, Folmer BJB, Meijer EW, Sijbesma RP (2001) Chem Rev 101:4071
20. Lehn J-M (2002) Polym Int 51:825
21. ten Cate AT, Sijbesma RP (2002) Macromol Rapid Commun 23:1094
22. Armstrong G, Buggy M (2005) J Mater Sci 40:547
23. Beijer FH, Sijbesma RP, Kooijman H, Spek AL, Meijer EW (1998) J Am Chem Soc 120:6761
24. Söntjens SHM, Sijbesma RP, van Genderen MHP, Meijer EW (2000) J Am Chem Soc 122:7487
25. Sijbesma RP, Meijer EW (2003) Chem Commun, p 5
26. Cates ME (1987) Macromolecules 20:2289
27. Cates ME, Candau SJ (1990) J Phys: Condens Matter 2:6869
28. Hanabusa K, Koto C, Kimura M, Shirai H, Kakehi A (1997) Chem Lett 429
29. Lightfoot MP, Mair FS, Pritchard RG, Warren JE (1999) Chem Commun, p 1945
30. Brunsveld L, Schenning APHJ, Broeren MAC, Janssen HM, Vekemans JAJM, Meijer EW (2000) Chem Lett, p 292
31. van Gorp JJ, Vekemans JAJM, Meijer EW (2002) J Am Chem Soc 124:14759
32. Ogata D, Shikata T, Hanabusa K (2004) J Phys Chem B 108:15503
33. Shikata T, Ogata D, Hanabusa K (2004) J Phys Chem B 108:508
34. Jang W-D, Aida T (2004) Macromolecules 37:7325
35. Hanabusa K, Kawakami A, Kimura M, Shirai H (1997) Chem Lett 191
36. Shikata T, Ogata D, Hanabusa K (2003) Nihon Reoroji Gakkaishi 31:229
37. Fan E, Yang J, Geib SJ, Stoner TC, Hopkins MD, Hamilton AD (1995) J Chem Soc, Chem Commun, p 1251
38. van Esch J, Schoonbeek F, de Loos M, Kooijman H, Spek AL, Kellogg RM, Feringa BL (1999) Chem Eur J 5:937
39. Boileau S, Bouteiller L, Lauprêtre F, Lortie F (2000) New J Chem 24:845
40. Bouteiller L, Colombani O, Lortie F, Terech P (2005) J Am Chem Soc 127:8893
41. Vonau F, Suhr D, Aubel D, Bouteiller L, Reiter G, Simon L (2005) Phys Rev Lett 94:066103
42. Pinault T, Isare B, Bouteiller L (2006) Chem Phys Chem 7:816
43. Ducouret G. private communication
44. Lortie F, Boileau S, Bouteiller L, Chassenieux C, Demé B, Ducouret G, Jalabert M, Lauprêtre F, Terech P (2002) Langmuir 18:7218

45. van der Gucht J, Besseling NAM, Knoben W, Bouteiller L, Cohen Stuart MA (2003) Phys Rev E 67:051106
46. Knoben W, Besseling NAM, Bouteiller L, Cohen Stuart MA (2005) Phys Chem Chem Phys 7:2390
47. Aggeli A, Bell M, Boden N, Keen JN, Knowles PF, McLeish TCB, Pitkeathly M, Radford SE (1997) Nature 386:259
48. Aggeli A, Bell M, Boden N, Keen JN, McLeish TCB, Nyrkova I, Radford SE, Semenov AN (1997) J Mater Chem 7:1135
49. Aggeli A, Nyrkova I, Bell M, Harding R, Carrick L, McLeish TCB, Semenov AN, Boden N (2001) Proc Natl Acad Sci USA 98:11857
50. Aggeli A, Bell M, Carrick LM, Fishwick CWG, Harding R, Mawer PJ, Radford SE, Strong AE, Boden N (2003) J Am Chem Soc 125:9619
51. Alig I, Braun D, Langendorf R, Wirth HO, Voigt M, Wendorff JH (1998) J Mater Chem 8:847
52. Boileau S, Bouteiller L, Foucat E, Lacoudre N (2002) J Mater Chem 12:195
53. Boils D, Perron M-E, Monchamp F, Duval H, Maris T, Wuest JD (2004) Macromolecules 37:7351
54. Lee C-M, Griffin AC (1997) Macromol Symp 117:281
55. Araki K, Takasawa R, Yoshikawa I (2001) Chem Commun, p 1826
56. Takasawa R, Murota K, Yoshikawa I, Araki K (2003) Macromol Rapid Commun 24:335
57. Castellano RK, Nuckolls C, Eichhorn SH, Wood MR, Lovinger AJ, Rebek J Jr (1999) Angew Chem Int Ed 38:2603
58. Castellano RK, Clark R, Craig SL, Nuckolls C, Rebek J Jr (2000) Proc Natl Acad Sci USA 97:12418
59. Gibson HW, Yamaguchi N, Jones JW (2003) J Am Chem Soc 125:3522
60. Hirschberg JHKK, Beijer FH, van Aert HA, Magusin PCMM, Sijbesma RP, Meijer EW (1999) Macromolecules 32:2696
61. Folmer BJB, Sijbesma RP, Versteegen RM, van der Rijt JAJ, Meijer EW (2000) Adv Mater 12:874
62. Armstrong G, Buggy M (2002) Polym Int 51:1219
63. Kato T, Mizoshita N, Kanie K (2001) Macromol Rapid Commun 22:797
64. Beginn U (2003) Prog Polym Sci 28:1049
65. Sivakova S, Rowan SJ (2003) Chem Commun, p 2428
66. Sivakova S, Wu J, Campo CJ, Mather PT, Rowan SJ (2006) Chem Eur J 12:446
67. Bladon P, Griffin AC (1993) Macromolecules 26:6604
68. Lee C-M, Jariwala CP, Griffin AC (1994) Polymer 35:4550
69. Kihara H, Kato T, Uryu T (1998) Liq Cryst 24:413
70. He C, Donald AM, Griffin AC, Waigh T, Windle AH (1998) J Polym Sci B 36:1617
71. He C, Lee C-M, Griffin AC, Bouteiller L, Lacoudre N, Boileau S, Fouquey C, Lehn J-M (1999) Mol Cryst Liq Cryst 332:251
72. Xu J, He C, Toh KC, Lu X (2002) Macromolecules 35:8846
73. Rogness DC, Riedel PJ, Sommer JR, Reed DF, Wiegel KN (2006) Liq Cryst 33:567
74. Lu X, He C, Griffin AC (2003) Macromolecules 36:5195
75. Malthête J, Levelut A-M, Liébert L (1992) Adv Mater 4:37
76. Albouy P-A, Guillon D, Heinrich B, Levelut A-M, Malthête J (1995) J Phys II France 5:1617
77. Pucci D, Veber M, Malthête J (1996) Liq Cryst 21:153
78. Allouchi H, Cotrait M, Malthête J (2001) Mol Cryst Liq Cryst 362:101
79. Garcia C, Malthête J (2002) Liq Cryst 29:1133

80. Ungar G, Abramic D, Percec V, Heck JA (1996) Liq Cryst 21:73
81. Percec V, Ahn C-H, Bera TK, Ungar G, Yeardley DJP (1999) Chem Eur J 5:1070
82. Percec V, Bera TK, Glodde M, Fu Q, Balagurusamy VSK (2003) Chem Eur J 9:921
83. Gearba RI, Lehmann M, Levin J, Ivanov DA, Koch MHJ, Barbera J, Debije MG, Piris J, Geerts YH (2003) Adv Mater 15:1614
84. Kishikawa K, Nakahara S, Nishikawa Y, Kohmoto S, Yamamoto M (2005) J Am Chem Soc 127:2565
85. Barbera J, Puig L, Romero P, Serrano JL, Sierra T (2005) J Am Chem Soc 127:458
86. Schmuck C, Wienand W (2001) Angew Chem Int Ed 40:4363
87. Jorgensen WL, Pranata J (1990) J Am Chem Soc 112:2008
88. Sartorius J, Schneider H-J (1996) Chem Eur J 2:1446
89. Lafitte VGH, Aliev AE, Horton PN, Hursthouse MB, Bala K, Golding P, Hailes HC (2006) J Am Chem Soc 128:6544
90. Lortie F, Boileau S, Bouteiller L (2003) Chem Eur J 9:3008
91. Williams DH, Gale TF, Bardsley B (1999) J Chem Soc, Perkin Trans 2, p 1331
92. Arnaud A, Bouteiller L (2004) Langmuir 20:6858
93. Hirschberg JHKK, Brunsveld L, Ramzi A, Vekemans JAJM, Sijbesma RP, Meijer EW (2000) Nature 407:167
94. Hirschberg JHKK, Koevoets RA, Sijbesma RP, Meijer EW (2003) Chem Eur J 9:4222
95. Castellano RK, Rudkevich DM, Rebek J Jr (1997) Proc Natl Acad Sci USA 94:7132
96. Castellano RK, Nuckolls C, Rebek J Jr (2000) Polym News 25:44
97. Chebotareva N, Bomans PHH, Frederik PM, Sommerdijk NAJM, Sijbesma RP (2005) Chem Commun, p 4967
98. Ashkenasy N, Horne WS, Ghadiri MR (2006) Small 2:99
99. Brunsveld L, Vekemans JAJM, Hirschberg JHKK, Sijbesma RP, Meijer EW (2002) Proc Natl Acad Sci USA 99:4977
100. Brunsveld L, Lohmeijer BGG, Vekemans JAJM, Meijer EW (2000) Chem Commun, p 2305
101. Fenniri H, Mathivanan P, Vidale KL, Sherman DM, Hallenga K, Wood KV, Stowell JG (2001) J Am Chem Soc 123:3854
102. Moralez JG, Raez J, Yamazaki T, Motkuri RK, Kovalenko A, Fenniri H (2005) J Am Chem Soc 127:8307
103. Fogleman EA, Yount WC, Xu J, Craig SL (2002) Angew Chem Int Ed 41:4026
104. Xu J, Fogleman EA, Craig SL (2004) Macromolecules 37:1863
105. Gothelf KV, Thomsen A, Nielsen M, Clo E, Brown RS (2004) J Am Chem Soc 126:1044
106. Waybright SM, Singleton CP, Wachter K, Murphy CJ, Bunz UHF (2001) J Am Chem Soc 123:1828
107. Abed S, Boileau S, Bouteiller L (2000) Macromolecules 33:8479
108. Folmer BJB, Sijbesma RP, Meijer EW (2001) J Am Chem Soc 123:2093
109. ten Cate AT, Kooijman H, Spek AL, Sijbesma RP, Meijer EW (2004) J Am Chem Soc 126:3801
110. Söntjens SHM, Sijbesma RP, van Genderen MHP, Meijer EW (2001) Macromolecules 34:3815
111. Scherman OA, Ligthart GBWL, Sijbesma RP, Meijer EW (2006) Angew Chem Int Ed 45:2072
112. Yamaguchi N, Gibson HW (1999) Chem Commun, p 789
113. Chen C-C, Dormidontova EE (2004) Macromolecules 37:3905
114. Ercolani G, Mandolini L, Mencarelli P, Roelens S (1993) J Am Chem Soc 115:3901
115. ten Cate AT, Dankers YW, Kooijman H, Spek AL, Sijbesma RP, Meijer EW (2003) J Am Chem Soc 125:6860

116. Folmer BJB, Sijbesma RP, Kooijman H, Spek AL, Meijer EW (1999) J Am Chem Soc 121:9001
117. Ohkawa H, Takayama A, Nakajima S, Nishide H (2006) Org Lett 8:2225
118. Ercolani G, Ioele M, Monti D (2001) New J Chem 25:783
119. Ercolani G (1998) J Chem Phys B 102:5699
120. Ercolani G (2003) J Chem Phys B 107:5052
121. Castellano RK, Rebek J Jr (1998) J Am Chem Soc 120:3657
122. Ligthart GBWL, Ohkawa H, Sijbesma RP, Meijer EW (2005) J Am Chem Soc 127:810
123. Berl V, Schmutz M, Krische MJ, Khoury RG, Lehn J-M (2002) Chem Eur J 8:1227
124. Lange RFM, van Gurp M, Meijer EW (1999) J Polym Sci Part A: Polym Chem 37:3657
125. Versteegen RM, van Beek DJM, Sijbesma RP, Vlassopoulos D, Fytas G, Meijer EW (2005) J Am Chem Soc 127:13862
126. Elkins CL, Viswanathan K, Long TE (2006) Macromolecules 39:3132
127. St. Pourcain CB, Griffin AC (1995) Macromolecules 28:4116
128. Lucas NL, van Esch J, Kellogg RM, Feringa BL (2001) Chem Commun, p 759
129. Takeshita M, Hayashi M, Kadota S, Mohammed KH, Yamato T (2005) Chem Commun, p 761
130. Yagai S, Iwashima T, Kishikawa K, Nakahara S, Karatsu T, Kitamura A (2006) Chem Eur J 12:3984
131. de Jong JJD, Lucas NL, Kellogg RM, van Esch JH, Feringa BL (2004) Science 304:278
132. Ikegami M, Ohshiro I, Arai T (2003) Chem Commun, p 1566
133. Folmer BJB, Cavini E, Sijbesma RP, Meijer EW (1998) Chem Commun, p 1847
134. Hofmeier H, El-ghayoury A, Schenning APHJ, Schubert US (2004) Chem Commun, p 318
135. Hofmeier H, Hoogenboom R, Wouters MEL, Schubert US (2005) J Am Chem Soc 127:2913
136. Kolomiets E, Lehn J-M (2005) Chem Commun, p 1519
137. Xu H, Hampe EM, Rudkevich DM (2003) Chem Commun, p 2828
138. Xu H, Rudkevich DM (2004) Chem Eur J 10:5432
139. Xu H, Rudkevich DM (2004) J Org Chem 69:8609
140. Xu H, Stampp SP, Rudkevich DM (2003) Org Lett 5:4583
141. Ishida Y, Aida T (2002) J Am Chem Soc 124:14017
142. Fishwick CWG, Beevers AJ, Carrick L, Whitehouse CD, Aggeli A, Boden N (2003) Nano Lett 3:1475
143. Bellesia G, Fedorov MV, Kuznetsov YA, Timoshenko EG (2005) J Chem Phys 122:134901
144. Nyrkova IA, Semenov AN, Aggeli A, Boden N (2000) Eur Phys J B 17:481
145. Nyrkova IA, Semenov AN, Aggeli A, Bell M, Boden N (2000) Eur Phys J B 17:499
146. Bellesia G, Fedorov MV, Timoshenko EG (2007) Physica A 373:455
147. Keller U, Müllen K, De Feyter S, De Schryver FC (1996) Adv Mater 8:490
148. Würthner F, Thalacker C, Sautter A (1999) Adv Mater 11:754
149. Würthner F, Thalacker C, Sautter A, Schärtl W, Ibach W, Hollricher O (2000) Chem Eur J 6:3871
150. Liu Y, Li Y, Jiang L, Gan H, Liu H, Li Y, Zhuang J, Lu F, Zhu D (2004) J Org Chem 69:9049
151. El-ghayoury A, Schenning APHJ, van Hal PA, van Duren JKJ, Janssen RAJ, Meijer EW (2001) Angew Chem Int Ed 40:3660
152. Varghese R, George SJ, Ajayaghosh A (2005) Chem Commun, p 593
153. Drain CM, Shi X, Milic T, Nifiatis F (2001) Chem Commun, p 287
154. Ercolani G (2001) Chem Commun, p 1416
155. Yamaguchi T, Ishii N, Tashiro K, Aida T (2003) J Am Chem Soc 125:13934

156. Yagai S, Higashi M, Karatsu T, Karatsu T, Kitamura A (2005) Chem Mater 17:4392
157. Korn ED, Carlier M-F, Pantaloni D (1987) Science 238:638
158. Cazacu A, Tong C, van der Lee A, Fyles TM, Barboiu M (2006) J Am Chem Soc 128:9541
159. Ikeda M, Nobori T, Schmutz M, Lehn J-M (2005) Chem Eur J 11:662
160. Sakamoto A, Ogata D, Shikata T, Urukawa O, Hanabusa K (2006) Polymer 47:956
161. Sakamoto A, Ogata D, Shikata T, Hanabusa K (2005) Macromolecules 38:8983
162. Yagai S, Iwashima T, Karatsu T, Karatsu T, Kitamura A (2004) Chem Commun, p 1114
163. Knoben W, Besseling NAM, Cohen Stuart MA (2006) Phys Rev Lett 97:068301
164. Lortie F, Boileau S, Bouteiller L, Chassenieux C, Lauprêtre F (2005) Macromolecules 38:5283
165. Knoben W, Besseling NAM, Cohen Stuart MA (2006) Macromolecules 39:2643
166. Hirschberg JHKK, Ramzi A, Sijbesma RP, Meijer EW (2003) Macromolecules 36:1429
167. Zou S, Schönherr H, Vancso GJ (2005) Angew Chem Int Ed 44:956
168. Kersey FR, Lee G, Marszalek P, Craig SL (2004) J Am Chem Soc 126:3038
169. Kim J, Liu Y, Ahn SJ, Zauscher S, Karty JM, Yamanaka Y, Craig SL (2005) Adv Mater 17:1749
170. van der Gucht J, Besseling NAM, Cohen Stuart MA (2002) J Am Chem Soc 124:6202
171. van der Gucht J, Besseling NAM, Fleer GJ (2003) J Chem Phys 119:8175
172. Clark TD, Ghadiri MR (1995) J Am Chem Soc 117:12364
173. Clark TD, Kobayashi K, Ghadiri MR (1999) Chem Eur J 5:782
174. Bassani DM, Darcos V, Mahony S, Desvergne JP (2000) J Am Chem Soc 122:8795
175. Masuda M, Jonkheijm P, Sijbesma RP, Meijer EW (2003) J Am Chem Soc 125:15935
176. Wilson AJ, Masuda M, Sijbesma RP, Meijer EW (2005) Angew Chem Int Ed 44:2275
177. Tang H, Sun J, Zhou X, Fu P, Xie P, Zhang R (2006) Macromol Chem Phys 204:155
178. Moreau JJE, Vellutini L, Wong Chi Man M, Bied C, Bantignies J-L, Dieudonné P, Sauvajol J-L (2001) J Am Chem Soc 123:7957
179. Moreau JJE, Pichon BP, Wong Chi Man M, Bied C, Pritzkow H, Bantignies J-L, Dieudonné P, Sauvajol J-L (2004) Angew Chem Int Ed 43:203
180. Martin RB (1996) Chem Rev 96:3043
181. Here K is defined as the equilibrium constant between two difunctional monomers. Alternatively, if the equilibrium constant between two functions (K') is considered (as in the example in Sect. 2.1.1), then one should slightly modify the relationship, because $K = 4K'$
182. Arnaud A, Belleney J, Boué F, Bouteiller L, Carrot G, Wintgens V (2004) Angew Chem Int Ed 43:1718
183. Oosawa F, Kasai M (1962) J Mol Biol 4:10
184. Oosawa F, Higashi S (1967) Prog Theor Biol 1:79
185. Simic V, Bouteiller L, Jalabert M (2003) J Am Chem Soc 125:13148
186. Ogawa K, Kobuke Y (2000) Angew Chem Int Ed 39:4070
187. Paulusse JMJ, Sijbesma RP (2003) Chem Commun, p 1494
188. Toupance T, Benoit H, Sarazin D, Simon J (1997) J Am Chem Soc 119:9191
189. Kolomiets E, Buhler E, Candau SJ, Lehn J-M (2006) Macromolecules 39:1173
190. Lopez D, Guenet J-M (2001) Macromolecules 34:1076
191. Abed S, Boileau S, Bouteiller L (2001) Polymer 42:8613
192. Shetty AS, Zhang J, Moore JS (1996) J Am Chem Soc 118:1019
193. Tobe Y, Utsumi N, Kawabata K, Nagano A, Adachi K, Araki S, Sonoda M, Hirose K, Naemura K (2002) J Am Chem Soc 124:5350
194. Hasegawa Y, Miyauchi M, Takashima Y, Yamaguchi H, Harada A (2005) Macromolecules 38:3724

195. Stoncius S, Orentas E, Butkus E, Ohrström L, Wendt OF, Wärnmark K (2006) J Am Chem Soc 128:8272
196. Sugawara N, Stevens ES, Bonora GM, Toniolo C (1980) J Am Chem Soc 102:7044
197. Cantrill SJ, Youn GJ, Stoddart JF (2001) J Org Chem 66:6857
198. Duweltz D, Lauprêtre F, Abed S, Bouteiller L, Boileau S (2003) Polymer 44:2295
199. Colombani O, Bouteiller L (2004) New J Chem 28:1373
200. Castellano RK, Craig SL, Nuckolls C, Rebek J Jr (2000) J Am Chem Soc 122:7876
201. Tellini VHS, Jover A, Garcia JC, Galantini L, Meijide F, Tato JV (2006) J Am Chem Soc 128:5728
202. Zeng H, Miller RS, Flowers RA, Gong B (2000) J Am Chem Soc 122:2635

Adv Polym Sci (2007) 207: 113–177
DOI 10.1007/12_2006_111

Published online: 13 January 2007

Supramolecular Materials Based On Hydrogen-Bonded Polymers

Gerrit ten Brinke[1] (✉) · Janne Ruokolainen[2] · Olli Ikkala[2]

[1]Laboratory of Polymer Chemistry, Materials Science Centre, University of Groningen, Nijenborgh 4, 9747 AG Groningen, The Netherlands
g.ten.brinke@rug.nl

[2]Department of Engineering Physics and Mathematics and Center for New Materials, Helsinki University of Technology, PO Box 2200, 02015 HUT Espoo, Finland

1	**Introduction**	115
1.1	Design Principles of Polymer-Based Hydrogen-Bonded Supramolecules	118
2	**Hydrogen-Bonded Block Copolymers**	118
2.1	Theoretical Considerations	118
2.2	Experimental Results	122
3	**Block Copolymer Blends with Hydrogen Bonding**	125
3.1	Archimedian Tiling	126
3.2	Dilute Solution	128
4	**Hydrogen-Bonded Comb Copolymers: Nonmesogenic Side Chains**	129
4.1	Homopolymer-Based Hydrogen-Bonded Comb Copolymers	130
4.1.1	Self-Assembly in the Bulk State	130
4.1.2	Self-Assembly in Dilute Solution: Hollow Spheres	134
4.2	Block Copolymer-Based Hydrogen-Bonded Comb Copolymers	137
4.2.1	Self-Assembly in the Bulk State: Two Length-Scale Structures	137
4.3	Applications	141
4.3.1	Nanoscale Protonic Polymeric Conductors	141
4.3.2	Switching Protonic Conductivity	142
4.3.3	Tridirectional Protonic Conductivity	143
4.3.4	Photonic Bandgap Materials	144
4.3.5	Exploiting the Thermoreversibility of Side Chain Bonding: Nanoporous Membranes	146
4.3.6	Exploiting the Thermoreversibility of Side Chain Bonding: Nano-Objects	150
4.4	Conjugated Polymer-Based Hydrogen-Bonded Comb Copolymers	152
5	**Hydrogen-Bonded Comb Copolymers: Mesogenic Side Chains**	157
5.1	Homopolymer-Based Hydrogen-Bonded Side-Chain Liquid-Crystalline Copolymers	157
5.2	Block Copolymer-Based Hydrogen-Bonded Side-Chain Liquid-Crystalline Copolymers	161
5.2.1	Temperature-Dependent Photonic Bandgap	161
5.2.2	AC Orientational Switching	162
6	**Layer-by-Layer Hydrogen Bonding Assembly**	164

7 **Hydrogen-Bonded Interpenetrating Polymer Networks: Reversible Volume Transitions** 167

8 **Conclusion and Outlook** 169

References 170

Abstract Combining supramolecular principles with block copolymer self-assembly offers unique possibilities to create materials with responsive and/or tunable properties. The present chapter focuses on supramolecular materials based on hydrogen bonding and (block co-) polymers. Several cases will be discussed where the self-assembled nanostructured morphology can be easily tuned using composition as the natural variable. A large body of the material reviewed concerns hydrogen-bonded side-chain (block co-) polymers. Side chains both with and without mesogenic units are discussed. Frequently the thermoreversibility of the hydrogen bonds allows for responsiveness of material properties to external stimuli such as temperature, pH, and electromagnetic fields. Temperature-dependent photonic bandgap, temperature-dependent proton conductivity, pH-erasable multilayers, temperature-induced volume transitions, and fast AC electric field-induced orientational switching of microdomains are the main examples.

Keywords Hydrogen bonding · Polymers · Block copolymers · Self-assembly · Tunability · Responsivity

Abbreviations

A	Adenine
bcc	Body-centered cubic
CPS	Monocarboxy-terminated polystyrene
CSA	Camphorsulfonic acid
DBSA	Dodecylbenzenesulfonic acid
DMF	Dimethylformamide
DNA	Deoxyribonucleic acid
FTIR	Fourier-transform infrared
HABA	2-(4-Hydroxybenzeneazo)benzoic acid
Hres	4-Hexylresorcinol
IPN	Interpenetrating polymer network
LBL	Layer-by-layer
LC	Liquid crystalline
LCST	Lower critical solution temperature
MSA	Methanesulfonic acid
NDP	Nonadecylphenol
ODT	Order–disorder transition
OG	Octyl gallate
PAA	Poly(acrylic acid)
PAAM	Poly(acrylamide)
PAE	Poly(amic acid)
PANI	Polyaniline
PB	Poly(1,4-butadiene)
PCEMA	Poly(2-cinnamoylethyl methacrylate)
PDP	Pentadecylphenol

PEI	Poly(ethylene imine)
PEK	Poly(ether ketone)
PEO	Poly(ethylene oxide)
PI	Polyimide
PIB	Poly(isobutylene)
PMAA	Poly(methacrylic acid)
PMMA	Poly(methyl methacrylate)
PNIPAM	Poly(*N*-isopropylacrylamide)
PPE	Poly(2,6-dimethyl-1,4-diphenyl oxide)
PPY	Poly(2,5-pyridine diyl)
PS	Polystyrene
PSOH	Hydroxyl-containing polystyrene
PSQ	Poly(methylsilsesquioxane)
P*t*BA	Poly(*tert*-butyl acrylate)
PTMSS	Poly(4-trimethylsilylstyrene)
PVA	Poly(vinyl alcohol)
PVPON	Poly(vinylpyrrolidone)
P2VP	Poly(2-vinylpyridine)
P4VP	Poly(4-vinylpyridine)
SAXS	Small-angle X-ray scattering
T	Thymine
TEM	Transmission electron microscopy
THF	Tetrahydrofuran
TSA	Toluenesulfonic acid
UPy	2-Ureido-4[1H]-pyrimidinones
χ	Flory–Huggins interaction parameter

1 Introduction

In soft matter block copolymers are a natural choice when aiming for nanostructures that are formed by a simple self-assembly process [1–10]. Several options exist to combine ordering at the nano level with tunable functional properties. In particular the combination of block copolymer self-assembly with supramolecular principles appears to be a very promising way to achieve complex responsive functions [11–21]. In supramolecular chemistry non-covalent intermolecular physical interactions, such as hydrogen-bonding, electrostatic, and donor–acceptor interactions and metal ion coordination, take the role of covalent bonds [11]. From these, hydrogen bonding arguably offers the best possibilities to create materials with properties that can be tuned by external stimuli. Hydrogen bonding is a donor–acceptor interaction specifically involving hydrogen atoms. The formation of a hydrogen bond is accompanied by a substantial gain in interaction energy, but at the same time there is a considerable loss in entropy due to the directional specificity of the bonding. As a consequence, hydrogen bonds are thermoreversible and display a high degree of cooperativity. This makes temperature one of the most obvious external triggers to tune material properties and several examples will be discussed. Other stimuli have been explored as well, including pH and photochemicals. In the latter case the modulation of the hydrogen-bonding interactions can be achieved, e.g., by azo compounds and their derivatives by employing the photochemically and thermally controlled reversible *trans*-to-*cis* isomerization [22, 23]. Additionally inhibitors for hydrogen bonding may be inserted, as for instance has been employed in the area of supramolecular polymers, where chain stoppers are added to control the average chain length [24]. Obviously, the control of the association–dissociation equilibrium is the key ingredient for the realization of responsive polymer materials based on hydrogen bonding. For tuning alone, hydrogen bonding offers additional opportunities. The composition of the system can be systematically varied using different relative amounts of the components involved. In this way self-assembled structures can be realized that have different characteristic length scale(s) and/or different structures. It is precisely this property that is used to great avail in the area of nano- and microporous films that are obtained by removing one of the components by dissolution from the self-assembled films [15, 17, 18, 25–27]. In one particular class of microporous films this is accomplished only after one of the components has become charged due to exposure to an aqueous solvent of a suitable pH [25–27].

Self-assembly in copolymer systems where the chemically different species are held together by hydrogen bonds has many features in common with ordinary copolymer systems provided the hydrogen bonds are strong enough. An often claimed advantage is that it is simpler to prepare hydrogen bonding-based samples than to synthesize the covalent analogues. Potentially more important is the fact that thermoreversibility enhances the equilibration process during self-assembly by releasing topological constraints that would have been present if the bonds had been permanent. Additionally, the presence of non-hydrogen-bonded species in many instances results in a significantly increased mobility, which is crucial in relation to the timescale associated with, e.g., electric field-induced switching of microdomain orientations [28].

The combination of reversibility, easy control of composition, and concurrent self-assembly behavior gives new opportunities for materials with tunable functional properties, which constitutes the guiding principle for the selection of material covered in this review. We found it useful to distinguish several classes of hydrogen-bonded polymer systems, excluding so-called supramolecular polymers where the chain character itself is due to hydrogen bonding. Our focus will be on systems involving hydrogen bonds with at least one of the components being a genuine polymer molecule from the very outset.

Self-assembly of block copolymer systems stands at the heart of nanotechnology with soft materials [1–10]. The covalent links between the chemically different blocks prevent macrophase separation which otherwise would result from the unfavorable interactions. Instead, self-assembly produces a variety of different nanostructures depending on molecular composition, molecular architecture, etc. Linear diblock copolymers are arguably the simplest representatives of this class of materials and a great effort has been put in during the last few decades to establish the complete phase diagram [29–37]. The direct generalization of this class of systems to systems involving hydrogen bonds are linear block copolymers formed by hydrogen bonding between chemically different telechelic polymers [38–42]. If the hydrogen bonding is strong enough, which as we will see generally requires the use of multiple hydrogen bonding, their self-assembly is similar but not identical to that of the corresponding covalent block copolymers. The added possibility of macrophase separation, e.g., between differently ordered microphase separated phases, is an essential new element. We will briefly discuss this class of systems, merely to elucidate the characteristic role of hydrogen bonding in relation to the self-assembled structures formed.

Then we will turn our attention to blends of block copolymers where the self-assembly is dominated by multiple hydrogen bonding between complementary blocks of the two constituents. The self-assembly of some of these systems has features in common with that of star copolymers and allows one to study the consequences of this architecture for the structures formed [43–45].

The most studied copolymer architectures are of the graft copolymer type, where relatively short side chains are attached to a main chain polymer backbone via hydrogen bonding [13]. Special attention will be given to hierarchically ordered systems that may be obtained if short side chains are attached to one of the blocks of a linear block copolymer [15, 17, 18, 28, 46]. The nature of the side chains can vary from simple flexible to additionally incorporating mesogenic units. Several examples will be shown where switching of functional properties can be induced by temperature and/or electromagnetic fields.

Chemically different polymers usually don't mix because the thermal motion of long-chain molecules is insufficient to overcome the generally unfavorable intermolecular interactions. There are many exceptions to this rule. Most of these are related to the intermolecular interactions being favorable, as may be the case in the presence of hydrogen bonding [47, 48]. Then miscible blends are obtained with the composition being the natural variable to tune material properties. A possibility to combine the properties of chemically different polymers that are not necessarily miscible is by the interpenetrating network concept, where both components form an intertwining network structure [49]. Although this does not allow for molecular miscibility, the phase separation is very local on a nanometer scale. If the interpenetrating network concept is combined with hydrogen bonding, thermoreversible hydrogels may be produced that exhibit volume transitions [50]. This is one of the clearest examples of the effect of the cooperativity of hydrogen bonding. Interpenetrating networks of complementary polymers form a condensed globular structure at low temperatures, with a quite sudden volume transition to a highly swollen state at elevated temperatures.

Finally, we will consider briefly the formation of multilayer thin films by layer-by-layer deposition of hydrogen-bonded polymer pairs [51, 52]. In this way a multilayer structure is obtained from potentially miscible polymer pairs. The stability of these films very much depends on the presence of hydrogen bonds, and pH may be used as an external trigger to erase the layered structure [53, 54] and selectively dissolve one of the components [25–27]. This procedure allows for the preparation of microporous films not unlike the nanoporous films obtained by dissolution of the hydrogen-bonded side groups from self-assembled block copolymer-based comb-shaped supramolecules [15, 17, 18].

The last two examples, thermoresponsible gels and layer-by-layer deposition, were first realized for ionic systems and only afterwards extended to hydrogen bonding [55–58].

1.1 Design Principles of Polymer-Based Hydrogen-Bonded Supramolecules

The different design principles used to create structures based on hydrogen bonding that will be covered in this review are illustrated in Fig. 1.

2 Hydrogen-Bonded Block Copolymers

Linear diblock copolymers formed by hydrogen bonding between two telechelic polymers have features in common with both polymer blends and diblock copolymer systems. For an excess of either component, the system resembles blend systems that are formed by blending a diblock copolymer with a homopolymer that is chemically identical to, and of the same length as, one of the blocks of the diblock copolymer. The latter systems have been extensively discussed in the past and a clear conclusion has been reached [1, 59–62]. The homopolymer is selectively solubilized in the corresponding microdomains; however, it does not significantly swell these domains. Rather it is segregated near the midplane, a situation that corresponds to the "dry brush" regime. If the hydrogen bonding is strong enough a similar behavior is expected for pseudo diblock copolymers in the strong segregation limit.

2.1 Theoretical Considerations

Let us start with a simple estimation of the hydrogen bonding strength required to avoid macrophase separation under strong segregation conditions, assuming that both telechelic polymers are present in the same number. Sup-

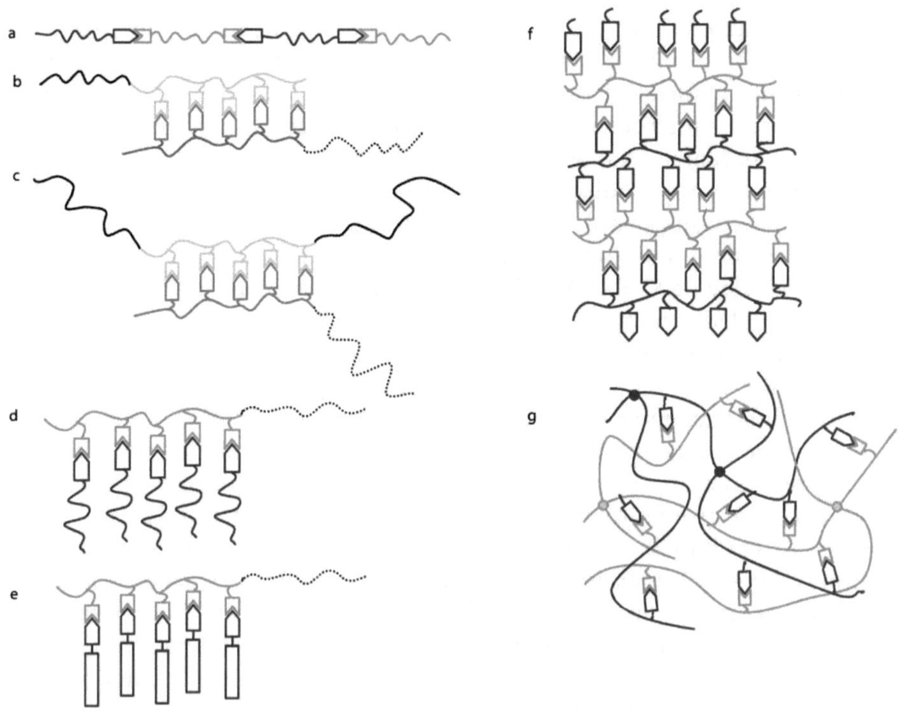

Fig. 1 **a** Hydrogen-bonded linear AB block copolymers. **b** Hydrogen-bonded AB/CD block copolymer blends: triblock architecture. **c** Hydrogen-bonded ABA/CD block copolymer blends: starlike architecture. **d** Hydrogen-bonded flexible side chain (block co-)polymers. **e** Hydrogen-bonded mesogenic side chain (block co-)polymers. **f** Layer-by-layer deposition based on hydrogen bonding. **g** Thermoresponsive hydrogels based on hydrogen bonding

pose the telechelic homopolymers P(A) and P(B) have equal length N and their interaction is as always described by the Flory–Huggins parameter χ satisfying $\chi N \gg 1$. In the case of genuine P(A)-*b*-P(B) diblock copolymers, a lamellar structure will be formed with a free energy per unit volume given by [63]

$$F/kT = \frac{3}{4N\nu}\left(\chi N\right)^{1/3} . \tag{1}$$

Here ν is the volume per "monomer", which is assumed to be equal for both species. In the case of pseudo diblock copolymers this free energy has to be offset by the free energy of the hydrogen bond ΔF_{hb}. Hence,

$$|\Delta F_{\mathrm{hb}}| > \frac{3}{2}\left(\chi N\right)^{1/3} kT . \tag{2}$$

Assuming characteristic values of $N = 1000$, $\chi = 0.1$, this implies $|\Delta F_{hb}| >$ 17.3 kJ/mol for $T = 298$ K. Such a value of the hydrogen bonding strength translates into association constants in the order of $K = 10^3\ M^{-1}$ as a minimum to observe microphase separation, much larger values being required for larger χ-parameter values.

For several examples discussed in this review the hydrogen-bonded complexes are based on phenol and pyridine groups. A study of the equilibrium constant in dilute solutions of phenol with pyridine in carbon tetrachloride gave a value of $K \cong 44\ M^{-1}$ [64]. This is far too small for the formation of self-assembled structures from pseudo linear block copolymers based on telechelic polymers end-functionalized with a single phenol and pyridine group. On the other hand, for pseudo graft copolymers formed by hydrogen bonding of alkyl phenols to, e.g., poly(4-vinylpyridine) (P4VP) it turns out to be more than sufficient due to the relatively short side chains in combination with a great many pyridine groups of P4VP. Hence, in the case of telechelic polymers, end groups that can form multiple hydrogen bonds have to be used to create diblock copolymer architectures. The same is true for the formation of *long* hydrogen-bonded supramolecular polymers, where the latter have been defined as "polymeric arrays of monomeric units that are brought together by reversible and highly directional secondary interactions.... The monomeric units of the supramolecular polymers themselves do not possess a repetition of chemical fragments" [24].

For blends of telechelic polymers forming linear pseudo diblock copolymers, the blend composition can be varied systematically using different amounts of the components. This has as a consequence that, except for a completely symmetric composition, there will always be an excess of nonassociated homopolymers of the majority component. As a consequence, near the phase boundaries between differently ordered mesophases a redistribution may occur in such a way that macrophase separation between the differently ordered mesophases takes place. To address these issues, a weak segregation theory of associating homopolymer blends was presented by Angerman et al. [65]. In this theoretical treatment the Landau free energy approach was adopted. Applied to genuine block copolymer systems the procedure involves the calculation of the minimization of the free energy with respect to the parameters characterizing the microstructures. In the present case there is the additional complication that the association equilibrium shifts due to the microphase separation. Hence, the free energy should be minimized simultaneously with respect to the association equilibrium, i.e., the composition of the blend in terms of the nonassociated homopolymers and the pseudo diblock copolymers, and the parameters characterizing the microstructure. One of the major accomplishments of Angerman was to demonstrate that, for arbitrary associating systems, this minimization procedure can actually be split into two independent steps. In the first step, the cluster composition is determined assuming the system to be homogeneous. In the second step, this composition is inserted into the expression for the Landau free energy without the nonlocal term. This latter term is an essential part of the Landau free energy expansion in the case of polydispersity [66–69]. In our case of associating telechelic homopolymers, the different clusters correspond to the homopolymers and the associated diblock copolymer. Since there are three different clusters (i.e., P(A), P(B), and P(A)-*block*-P(B)) and only two chemically different species, the nonlocal term is essential. However, as Angerman demonstrated, the error made in neglecting the change in cluster composition due to the formation of a microstructure exactly cancels the error made due to the omission of the nonlocal term. As a result, the calculation of the phase diagram becomes relatively straightforward, especially for the case under consideration. Figure 2 presents a typical phase diagram for pseudo diblock copolymers based on hydrogen bonding between two homopolymers of equal length. The parameters have been selected in such a way that at elevated temperatures a closed loop macrophase separation regime is found. The presence of the closed loop region is directly related to the directionality of the hydrogen bonds; at elevated temperature the number of hydrogen bonds is strongly reduced due to the entropy penalty associated with such a bond. As a consequence, the dispersion forces dominate and macrophase separation may occur [70].

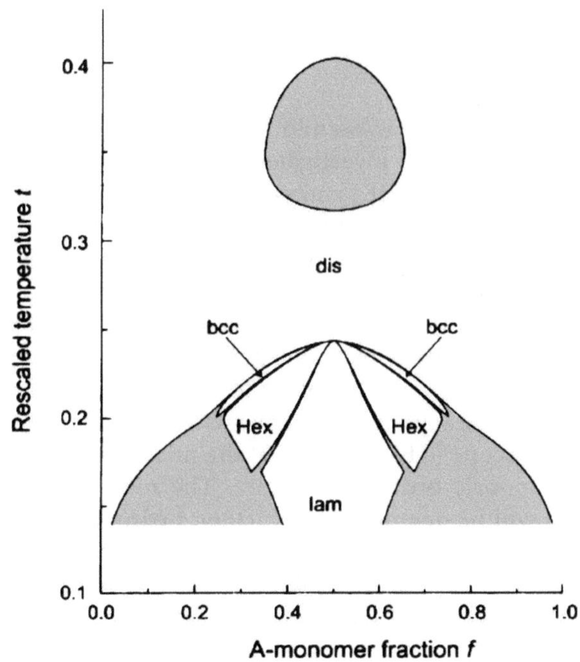

Fig. 2 Phase diagram of P(A)/P(B) homopolymer blend, where P(A) and P(B) can form a pseudo diblock copolymer due to hydrogen bonding [65]

The case $f = 0.5$ corresponds to an equal number of P(A) and P(B) telechelic homopolymers. In this case, below a critical temperature a lamellar structure is always formed with the nonassociated homopolymers all incorporated in the corresponding layers. For an excess of either component hexagonally ordered cylinders as well as bcc ordered spheres may be obtained, but only as long as the segregation is not too strong. At stronger segregation where, however, the Landau approach starts to become less accurate, the phase diagram is dominated by macrophase separation involving a microphase separated layered structure and an isotropic phase. In this regime the behavior is similar to that of blend systems formed by blending a diblock copolymer with a homopolymer that is chemically identical to, and of the same length as, one of the blocks of the diblock copolymer. For the particular example presented in Fig. 2, the closed loop macrophase separated regime is separated from the microphase separated regime and the self-assembled system becomes a homogeneous mixed system above the order–disorder temperature. If the hydrogen bonding is less strong, the self-assembled state may directly transform into a macrophase separated system on heating. The discussion given in [65] was restricted to the classical morphologies of pseudo diblock copolymers. It was subsequently extended to pseudo triblock copolymers, considering in addition other structures such as bicontinuous morphologies [71].

2.2 Experimental Results

Experimental studies on linear pseudo block copolymers are still relatively scarce. Binder and coworkers investigated the self-assembly of the telechelic polymers poly(isobutylene) (PIB) and poly(ether ketone) (PEK) [39, 40]. Two different hydrogen bonding systems were employed: thymine/triazine and Hamilton receptor/barbituric acid. Scheme 1 illustrates these systems. The authors restricted their investigations to 1 : 1 molar ratios of both components. For the weaker thymine/triazine association the samples cast from a common solvent showed a microphase separated, presumably lamellar, structure up to ca. 150 °C corresponding to the T_g of PEK. For the stronger bonding the microphase separated state persisted up to ca. 230 °C. Transmission electron microscopy (TEM) images of the microphase separated systems revealed a rather poorly ordered structure. The reason is not immediately obvious since, as will be demonstrated further, hydrogen-bonded side-chain polymers usually self-assemble into well-ordered structures. In a very recent study by Matsushita and coworkers [72], blends were prepared consisting of poly(4-trimethylsilylstyrene) (PTMSS; $M_n = 11\,000$, $M_w/M_n = 1.03$) and deuterated polystyrene (PS_{d8}; $M_n = 10\,000$, $M_w/M_n = 1.03$) end-decorated with complementary oligonucleotides, i.e., thymidine phosphates and deoxyadenosine phosphates. Also in this case TEM and SAXS revealed a poorly

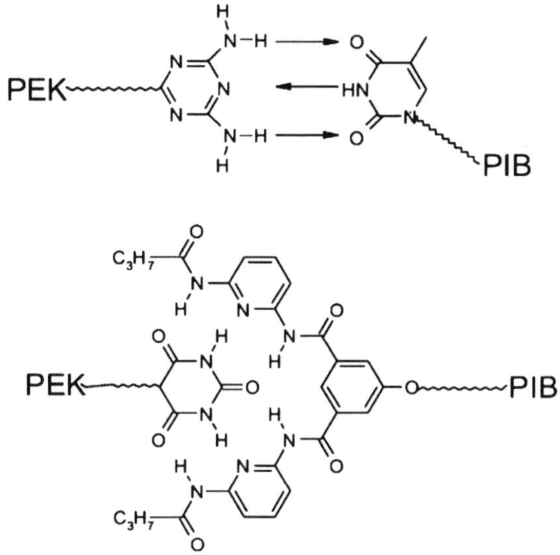

Scheme 1 Thymine/triazine and Hamilton receptor/barbituric acid interactions with association constants of $\sim 800\ \mathrm{M}^{-1}$ and $\sim 3 \times 10^4\ \mathrm{M}^{-1}$, respectively [42]

ordered microphase separated morphology as illustrated in Fig. 3. The same figure also shows a TEM image of an ionically bonded diblock copolymer obtained from mono-end-aminated polyisoprene ($M_n = 14\,000$) and di-end-sulfonated polystyrene ($M_n = 14\,000$) [73]. In contrast to the hydrogen-bonded examples, here a very well-ordered lamellar morphology is obtained. This system is additionally of interest due to the giant thermal tunability of the lamellar spacing from 20 to 55 nm on going from 110 to 190 °C.

Most of the research in the area of hydrogen-bonded polymers in the group of Meijer concerns supramolecular polymers consisting of monomeric

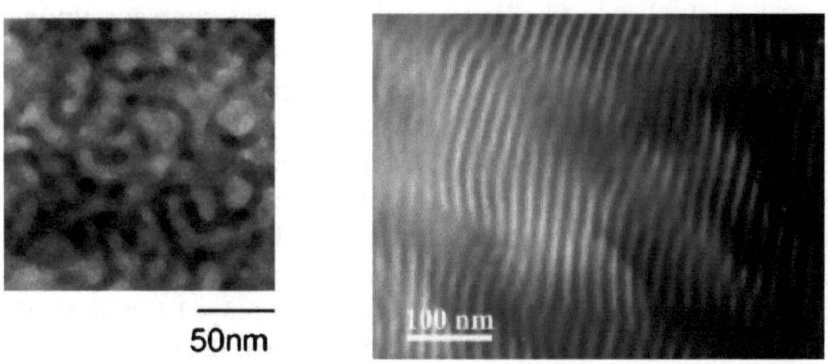

Fig. 3 TEM images. *Left*: 1 : 1 molar ratio blend of poly(4-trimethylsilylstyrene) and deuterated polystyrene end-decorated with complementary oligonucleotides. Reprinted with permission from [72]. © 2006 American Chemical Society. *Right*: blend of 60 wt % mono-end-aminated polyisoprene with di-end-sulfonated polystyrene [73]

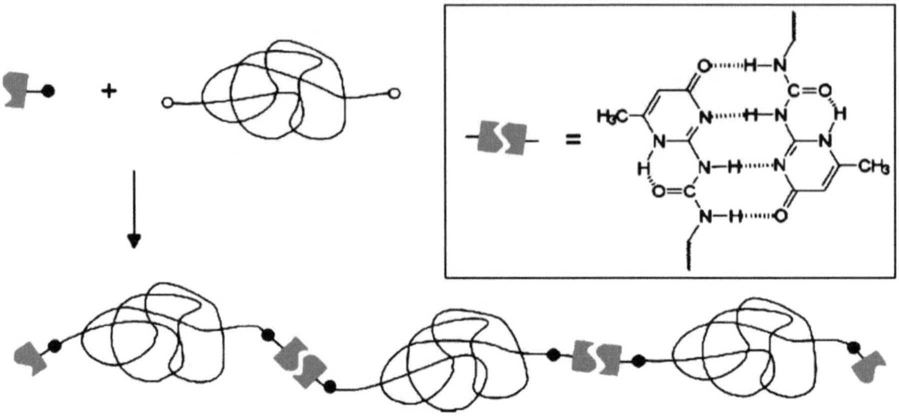

Fig. 4 Supramolecular polymers due to quadruple hydrogen bonding between bifunctional ureidopyrimidinone derivatives [74]

units that are brought together by reversible and highly directional secondary interactions [24]. However, recently the group also investigated combinations of monomers and polymers [74, 75]. The focus is on exploiting the strong dimerization of 2-ureido-4[1H]-pyrimidinones (UPy) by quadruple hydrogen bonding. Telechelic poly(ethylene/butylene) with UPy end groups was synthesized and a large chain extension, as illustrated in Fig. 4, was achieved as verified by dynamical–mechanical thermal analysis. This application is interesting from a mechanical properties point of view. As far as microphase separated structures are concerned the following example from the same group is of interest. Rigid rodlike segments were formed based on the self-complementary quadruple hydrogen bonding unit ureidotriazine with trialkoxyphenyl substituents. Copolymers were obtained by combining these with poly(ethylene/butylene) functionalized with either one or two ureidotriazine units. The authors observed that phase separation between the components resulted in a fibrillar morphology that coarsens upon annealing [75]. However, the nature of the phase separation is far from clear.

All these examples involved multiple hydrogen bonding units to obtain a microphase separated morphology based on end group association between polymer chains. The bonding strength required depends strongly on the chain lengths involved and one example is known where a simple hydrogen bond suffices [76]. It concerns tertiary amino-terminated poly(methyl methacrylate) (PMMA) and dodecylbenzenesulfonic acid (DBSA) (Scheme 2). The polymer employed had a molar mass of $M_n = 6.3 \times 10^3$ g/mol, $M_w/M_n = 1.19$. For a weight fraction of DBSA in between ca. 0.25 and 0.7, a lamellar structure is formed, with a composition-dependent order–disorder transition in the range of 40 to 120 °C. For other compositions a homogeneous mixture is formed. Clearly, there is considerable scope for further investigations, and we should expect to see many more results in the near future.

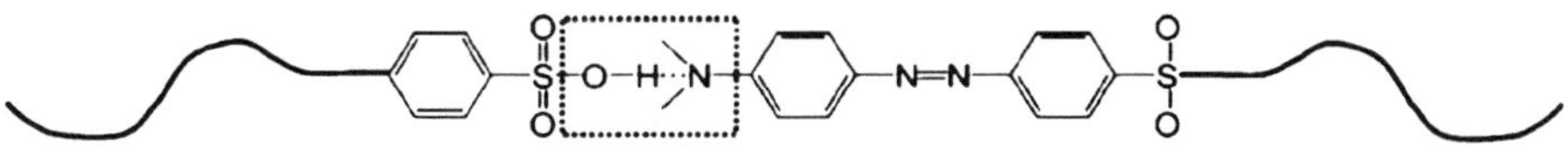

Scheme 2 Hydrogen bonding between DBSA and tertiary amino-terminated poly(methyl methacrylate) [76]

3
Block Copolymer Blends with Hydrogen Bonding

Having considered hydrogen bonding between telechelic polymers leading to linear block copolymer architectures, we consider next block copolymer blends involving multiple hydrogen bonds between complementary blocks. Miscible polymer blends based on hydrogen bonding between complementary polymers have been extensively discussed in the literature [47, 48]. The extension to block copolymer blends is of a much more recent date. The most interesting studies to date deal with the complex morphologies resulting from binary mixtures of block copolymers, where the different block copolymers contain complementary blocks that are miscible due to hydrogen bonding.

The group of Abetz [77] used polystyrene-*block*-poly(1,2-butadiene)-*block*-poly(*tert*-butyl methacrylate) triblock copolymers with partially hydrolyzed poly(*tert*-butyl methacrylate) blocks, together with polystyrene-

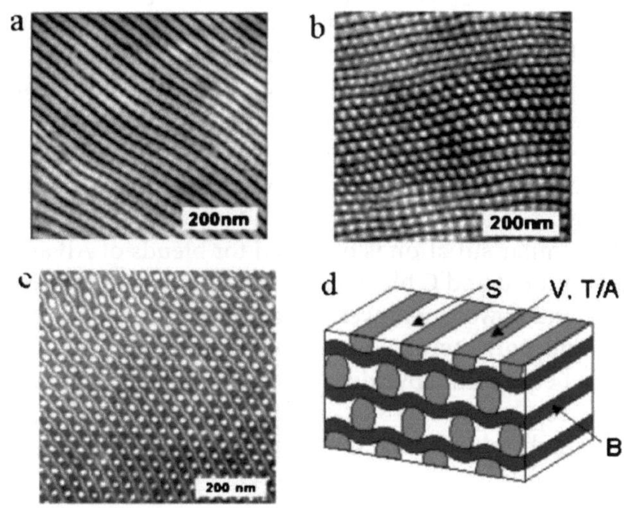

Fig. 5 TEM micrographs of a polystyrene-*block*-poly(1,2-butadiene)-*block*-poly(*tert*-butyl methacrylate) triblock copolymer that is 18% hydrolyzed and polystyrene-*block*-poly(2-vinylpyridine) diblock copolymer. S denotes polystyrene, V denotes poly(2-vinylpyridine), T/A denotes the partially (18%) hydrolyzed poly(*tert*-butyl methacrylate), and B denotes poly(1,2-butadiene). Reprinted with permission from [77]. © 2003 American Chemical Society

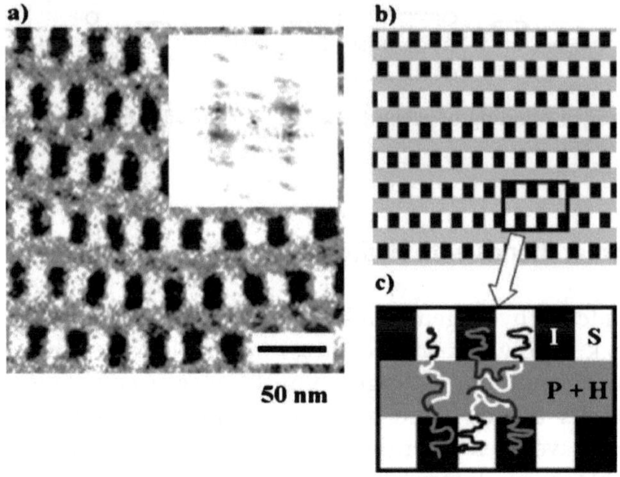

Fig. 6 TEM image of a 50 : 50 diblock copolymer blend of poly(isoprene-*block*-2-vinylpyridine) (IP) and poly(styrene-*block*-4-hydroxystyrene) (SH). Reprinted with permission from [44]. © 2005 American Chemical Society

block-poly(2-vinylpyridine) (PS-*b*-P2VP) or poly(2-vinylpyridine)-*block*-poly(cyclohexyl methacrylate) diblock copolymers. It turned out that full hydrolysis resulted in very strong hydrogen bonding between the methacrylic acid groups and the pyridine groups, effectively suppressing the formation of ordered superlattices. For moderate hydrolysis levels interesting superlattices were found. As a characteristic example, Fig. 5 illustrates the structure found for a blend of PS-*b*-P2VP and the triblock copolymer hydrolyzed up to 18% degree. Due to the hydrogen bonds between the methacrylic acid and the 2-vinylpyridine groups, the corresponding blocks segregate together. Since both components contain a styrene block we are effectively dealing with three different domains.

A somewhat similar situation is obtained for blends of AB and CD diblock copolymers, with the B and C blocks strongly interacting via hydrogen bonding. Matsushita and coworkers investigated blends of poly(isoprene-*block*-2-vinylpyridine) and poly(styrene-*block*-hydroxystyrene) [44]. The strong hydrogen bonding between the pyridine groups and the hydroxystyrene groups again forces the corresponding blocks together, thus creating self-assembled three-phase hierarchical structures. An example is given in Fig. 6.

3.1 Archimedian Tiling

In a subsequent paper the AB diblock copolymer was replaced by a BAB triblock copolymer. Poly(2-vinylpyridine-*block*-isoprene-*block*-2-vinylpyridine) with molar ratio 0.93 : 0.07 between isoprene and vinylpyridine was blended with poly(styrene-*block*-4-hydroxystyrene) with a molar ratio of 0.86 : 0.14 between styrene and 4-hydroxystyrene [45]. Again the poly(4-hydroxystyrene) blocks and the poly(2-vinylpyridine) blocks segregate together due to hydrogen bonding. The self-assembled structures found for these blends are shown to correspond to so-called ($3^3.4^2$) and (3.4.6.4) Archimedian tiling patterns. There are 11 different Archimedian tilings (see Fig. 7), which by definition consist of a tessellation of regular polygons where all vertices are of the same type [78]. A tiling is denoted by its vertex type such that, for instance, (3.12^2) denotes the tiling where one triangle and two 12-gons meet at every vertex. Archimedian tilings were first discussed in relation to ABC star polymers and it was argued that only three patterns, (6^3), (4.8^2), and (4.6.12), are allowed since only three polygons should meet in a vertex and only even polygons should appear [43]. However, this classification was based on regarding polygons as polymeric domains directly tiled. Subsequently, Matsushita and coworkers extended the concept by employing tiling as a skeleton and demonstrated the existence of the complex Archimedian tiling ($3^2.4.3.4$) for a star-branched polymer composed of polyisoprene, polystyrene, and poly(2-vinylpyridine). Figure 8 illustrates the "skeleton" concept for a 2 : 1 blend of poly(2-vinylpyridine-*block*-isoprene-*block*-2-vinylpyridine) and poly(styrene-*block*-4-hydroxystyrene) with a microdomain assembly

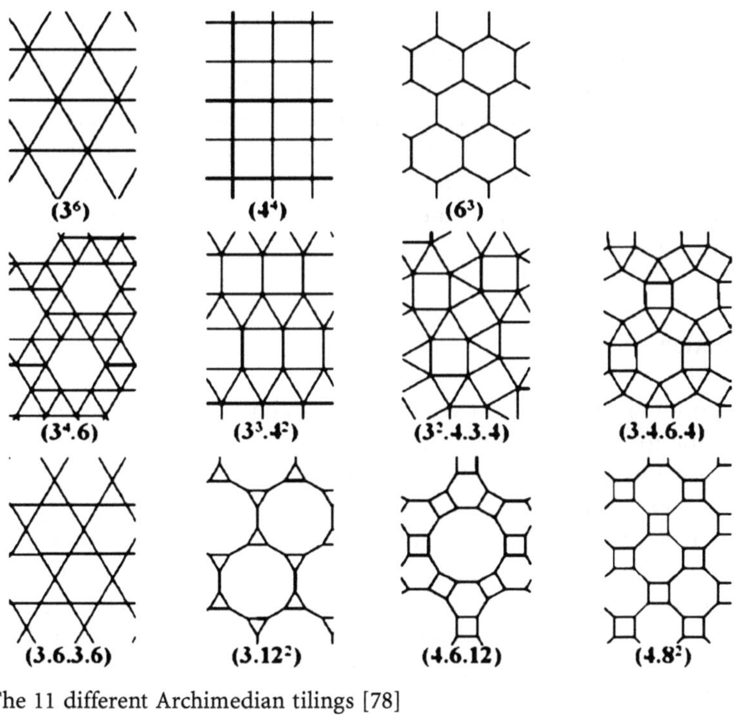

Fig. 7 The 11 different Archimedian tilings [78]

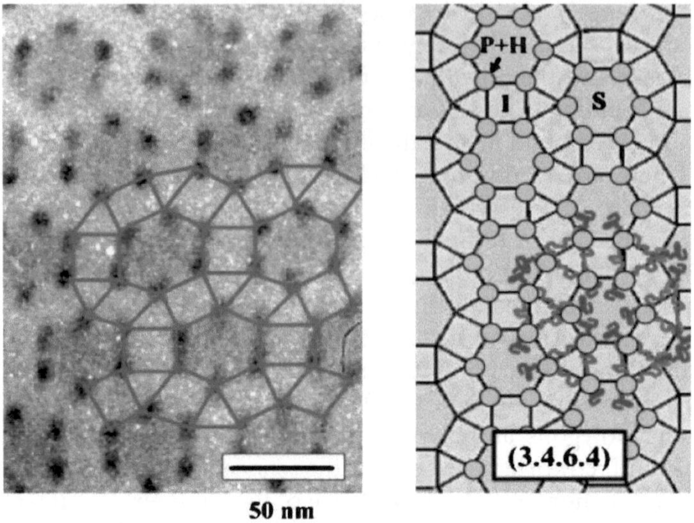

Fig. 8 Illustration of a (3.4.6.4) Archimedian tiling self-assembled structure of a 2 : 1 blend of poly(2-vinylpyridine)-*block*-polyisoprene-*block*-poly(2-vinylpyridine) and polystyrene-*block*-poly(4-hydroxystyrene). P denotes poly(2-vinylpyridine), H denotes poly(4-hydroxystyrene), S denotes polystyrene and I denotes polyisoprene. Reprinted with permission from [45]. © 2006 American Chemical Society

corresponding to a (3.4.6.4) Archimedian tiling. For a 1 : 1 blend composition on the other hand, a $(3^3.4^2)$ Archimedian tiling pattern is found [43].

These examples demonstrate that blending of block copolymers involving intermolecular association via hydrogen bonding is a very promising route toward hierarchical two-dimensional nanostructures. As in the case of pseudo linear block copolymers, tuning in these types of materials merely involves the possibility to steer the microphase separated morphology by selecting the composition.

3.2
Dilute Solution

Block copolymer micelles and vesicles have been studied for several decades and continue to attract a lot of attention [79–88]. They are usually formed in selective solvents, which dissolve only one of the blocks. Self-assembly gives rise to the formation of nanoparticles with a variety of shapes ranging from spheres to tubes to donuts. Hydrogen bonding between complementary block copolymers as an additional tool to create even more morphologies has been introduced only recently. Liu and coworkers [89, 90] studied the coaggregation of poly(*tert*-butyl acrylate)-*block*-poly(2-cinnamoyloxyethyl methacrylate) (P*t*BA-*b*-PCEMA) and polystyrene-*block*-poly(2-cinnamoyloxyethyl methacrylate) (PS-*b*-PCEMA) with the PCEMA blocks of the different diblock copolymers tagged by the hydrogen-bonding DNA base pair thymine and adenine, respectively. Hydrogen bonding between the associating blocks led to interesting block copolymer aggregation behavior and morphologies. Even more recently Zhou and coworkers [91] used the hydrogen bonding between poly(acrylic acid) and the poly(ethylene oxide) block of a poly(ethylene oxide)-*block*-polybutadiene diblock copolymer to demonstrate controllable vesicle formation.

4 Hydrogen-Bonded Comb Copolymers: Nonmesogenic Side Chains

By far the most studied class of hydrogen-bonded copolymer systems involves the comb-shaped architecture where side groups are attached to a linear polymer by hydrogen bonding [13]. This concept was first applied in the area of liquid-crystalline materials where supramolecular side-chain liquid-crystalline polymeric materials were produced by hydrogen bonding mesogenic side groups to a polymeric backbone [92]. We will address this in more detail in the next section, but start by considering the simplest case obtained by hydrogen bonding relatively short flexible-chain molecules to a polymeric backbone. Other types of interactions have been employed frequently during the last decade as well. Besides coordination [93, 94], especially ionic bonding [95–103], leading to so-called polyelectrolyte–surfactant complexes, has received a lot of attention.

Comb-shaped supramolecules obtained by hydrogen bonding have a characteristic amphiphilic nature. They usually consist of a polar "backbone structure", comprising the polymer backbone and the polar head of the surfactant molecule, and nonpolar (alkyl) side chain material. In most cases stoichiometric complexation, i.e., one side chain per backbone monomer, is considered, but it is one of the advantages of hydrogen bonding that it can be varied systematically without additional effort. Stoichiometric conditions correspond on the one hand to a very high "grafting density", but on the other hand, the volume fraction of the nonpolar alkyl tails may well at the same time be of the same order of magnitude as the volume fraction of the polar part. Hence, interesting microphase separated morphologies might be expected and are indeed observed. Another important aspect is that, since the hydrogen bonds usually are not too strong, a number of unoccupied hydrogen bonding sites of the backbone may be located inside the polar domains without severe energetic penalties. Hence, highly ordered simple structures, such as well-defined layers, should be the rule in hydrogen-bonded comb-shaped supramolecule systems. In the case of ionic bonding in polyelectrolyte–surfactant complexes, the strong side-chain bonding puts severe constraints on the structures, thus leading to all kinds of modulated structures [104, 105].

4.1
Homopolymer-Based Hydrogen-Bonded Comb Copolymers

4.1.1
Self-Assembly in the Bulk State

The study of the bulk state properties of systems consisting of comb-shaped supramolecules obtained by hydrogen bonding of short, flexible, nonmesogenic side chains was initiated by the groups of Ikkala and ten Brinke [15, 106–110]. In their work P4VP hydrogen bonded to alkyl phenols, notably pentadecylphenol (PDP) and nonadecylphenol (NDP), acts as a typical prototype system (Scheme 3). Self-assembly of hydrogen-bonded comb-shaped supramolecules in the melt state was studied in detail for PDP hydrogen bonded to P4VP. Infrared measurements demonstrated that below ca. 100 °C, hydrogen bonding is fairly complete for an amount of PDP not exceeding the stoichiometric composition (i.e., one PDP molecule for every pyridine group) [108, 111]. Hence, for a stoichiometric composition, denoted as $P4VP(PDP)_{1.0}$, most monomer units carry a PDP side chain. Although this corresponds to a high grafting density, it does not in this case imply that the "composition" of the comb molecule is very asymmetric. The segregating species are the nonpolar alkyl tails on the one hand, and the more polar P4VP–phenol complex on the other. In these terms the composition of $P4VP(PDP)_{1.0}$, assuming approximately all hydrogen bonds are formed, is close to 50 w/w %. Temperature-dependent small-angle X-ray scattering (SAXS) data show that the system orders below the order–disorder transition (ODT) at ca. 65 °C into a lamellar morphology [109, 110]. A typical SAXS pattern for a slightly smaller amount of PDP, i.e., $P4VP(PDP)_{0.85}$, is presented in Fig. 9. In the melt state the second-order peak is only very faintly present. This confirms the nearly 50 : 50 composition of the segregating species, as the thickness of the alternating layers has to be very similar to extinguish the second-order reflection. The third-order reflection is prominently present; however, it is located outside the q range observed with the experimental setup. Above the ODT a characteristic correlation hole peak is present. The position and height of this peak are strongly influenced by the fraction of hydrogen bonds. On heating this fraction will decrease, and the peak height will also de-

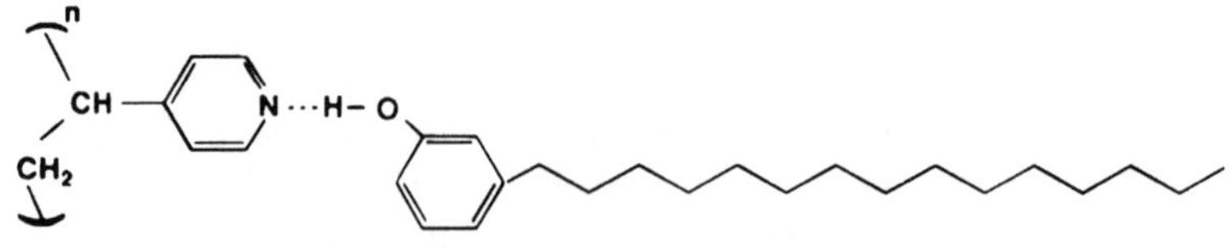

Scheme 3 Stoichiometric $P4VP(PDP)_{1.0}$ comb-shaped supramolecule [106]

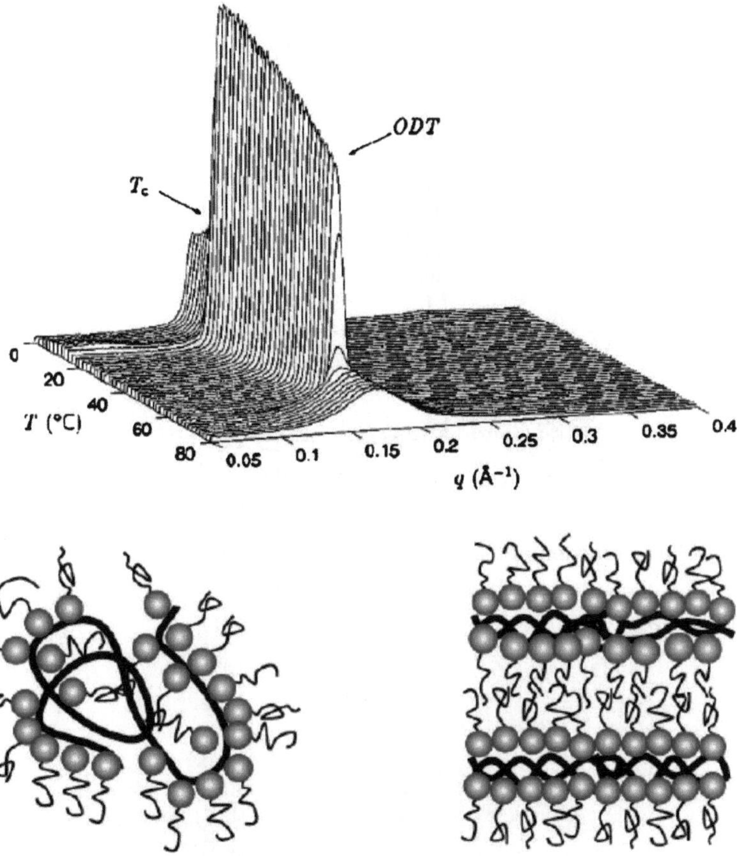

Fig. 9 Temperature-dependent SAXS of P4VP(PDP)$_{0.85}$ demonstrating order–disorder transition at ca. 65 °C and cartoon of corresponding lamellar self-assembly [109]

crease and gradually shift to smaller angles. Theoretical modeling indicates that in the order of 70% of the hydrogen bonds have to be broken before the correlation hole completely vanishes [112]. Around room temperature the alkyl tails crystallize in an interdigitated fashion [113]. The crystallization destroys the symmetry somewhat and the second-order peak becomes more clearly visible, in particular if an excess of PDP is used, as demonstrated for P4VP(PDP)$_{1.5}$ in Fig. 10. The lamellar morphology, concluded from the SAXS data, is corroborated by TEM (Fig. 11) using instead of PDP the slightly larger nonadecylphenol (NDP) [114]. In contrast, no layered self-assembly was observed for stoichiometric mixtures of P4VP and alkyl carboxylic acids (dodecanoic, hexadecanoic, and nonadecanoic). SAXS showed a clear correlation hole peak, indicating that a substantial number of hydrogen-bonded side chains are formed, but no ODT occurred upon cooling [115]. Rather, macrophase separation took place due to crystalliza-

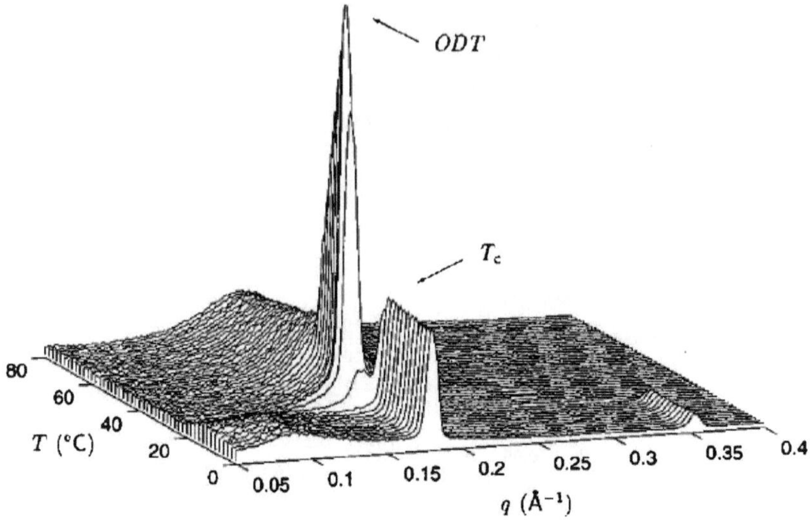

Fig. 10 SAXS as a function of temperature for $P4VP(PDP)_{1.5}$ [109]

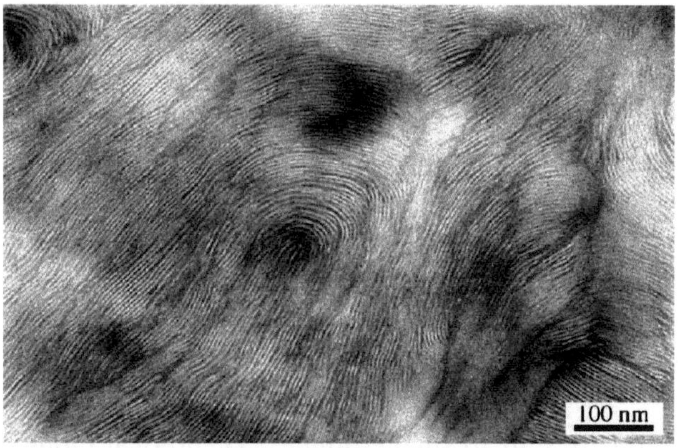

Fig. 11 TEM image of the layered structure of $P4VP(NDP)_{1.0}$ [114]

tion of the alkyl carboxylic acids. However, when a stronger base, imidazole ($pK_b = 8.75$ compared to 7.05 for pyridine) was selected self-assembled layered structures were found for several poly(1-vinylimidazole)–alkanoic acid complexes [116].

Chen and coworkers studied the self-assembly of comb-shaped supramolecules obtained by poly(ethylene oxide) (PEO) and DBSA [117]. Due to the hydrogen bonding between PEO and DBSA hydrogen-bonded side-chain polymers were again formed and, as in the case of P4VP and alkyl phenols, the repulsion happened to be strong enough to induce self-assembly into a lamellar morphology for 0.5 or 1.0 DBSA molecule per PEO repeat. The transition to the ordered state took place at ca. 60 and 70 °C, respectively.

The phase behavior of systems like P4VP(PDP) and PEO(DBSA) is relatively straightforward. The hydrogen bonding groups present the (strong) attraction between both species. The repulsion, which induces the layered structure, is due to the unfavorable polar–nonpolar interactions. A more complex behavior has been observed for the same polymer P4VP when the pyridine groups are first protonated with methanesulfonic acid (MSA) and then hydrogen bonded to, e.g., PDP. The combination of protonation and hydrogen bonding is of particular interest in the area of conjugated polymers, which will be discussed further on. PDP still hydrogen bonds to the stoichiometric P4VP(MSA)$_{1.0}$ salt and a comb-shaped supramolecule is formed. Scheme 4 illustrates the complex. The system with MSA undergoes an order–disorder transition to a layered structure at ca. 100–120 °C. Compared with the ca. 65 °C for P4VP(PDP)$_{1.0}$, this implies a somewhat larger repulsion in the P4VP(MSA)$_{1.0}$(PDP)$_{1.0}$ system. At elevated temperatures the system goes through an interesting sequence of transitions. At 170 °C the PDP first macrophase separates from the P4VP(MSA)$_{1.0}$ complex, only to homogeneously mix again at ca. 195 °C. Finally, at even higher temperatures, the system macrophase separates again [46]. This closed loop (re-entrant) phase behavior is quite characteristic for hydrogen bonding mixtures [70]. In this case it indicates the presence of a somewhat weaker hydrogen bonding combined with a somewhat stronger repulsion between the two species P4VP(MSA) and PDP as compared to P4VP and PDP. Figure 12 demonstrates this behavior by optical microscopy and SAXS. The latter clearly reveals the ODT around 100 °C. Above the ODT, the height of the correlation peak decreases and it shifts to smaller angles. This is a clear signal of a strong decrease in the number of hydrogen bonds. Around 170 °C nearly all hydrogen bonds appear to be broken and macrophase separation between PDP and P4VP(MSA)$_{1.0}$ sets in. The SAXS data show the corresponding strong scattering in the forward direction.

If instead of MSA toluenesulfonic acid (TSA) is used, the ODT is considerably higher at ca. 220 °C, indicating still stronger repulsion. Moreover,

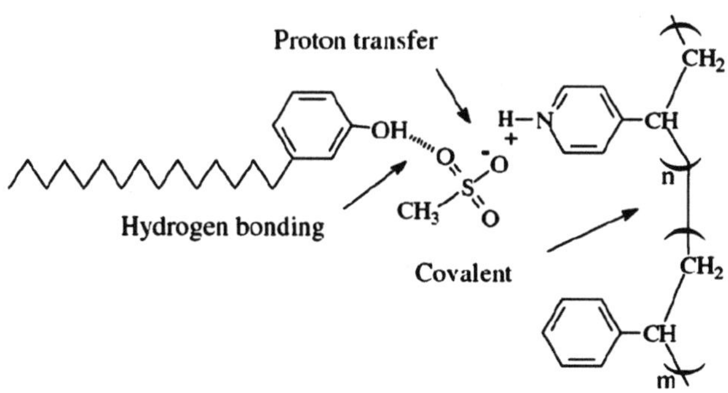

Scheme 4 PS-*b*-P4VP(MSA)$_{1.0}$ (PDP)$_{1.0}$ comb-shaped supramolecules [46]

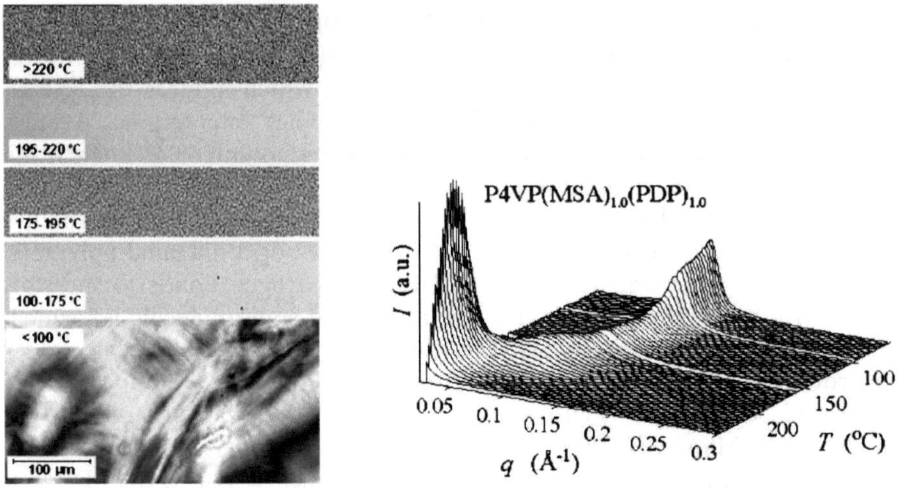

Fig. 12 SAXS as a function of temperature and sequence of optical microscopy pictures of P4VP$(MSA)_{1.0}(PDP)_{1.0}$ at indicated temperatures [46]

at elevated temperatures macrophase separation is not observed. Hence, this stronger repulsion must be accompanied by stronger hydrogen bonding, possibly due to additional phenyl ring stacking [118]. The polysalts P4VP$(MSA)_{1.0}$ and P4VP$(TSA)_{1.0}$ are examples of acid–base complexes studied extensively during the last few decades in relation to solid-state ionics, but before we address this issue, we will first turn our attention briefly to structure formation of hydrogen-bonded side-chain polymers in solution.

4.1.2 Self-Assembly in Dilute Solution: Hollow Spheres

The self-assembly of blends of hydrogen-bonding block copolymers in dilute solution has been addressed briefly in Sect. 3.2. A block copolymer-free strategy was introduced by Jiang and coworkers [119–127]. They used monocarboxy-terminated polystyrene (CPS) oligomers together with P4VP to prepare hydrogen-bonded graft copolymers in a common solvent chloroform. Then the soluble complex may self-assemble into a stable micelle after a selective solvent, e.g., toluene for CPS, is added. Figure 13 illustrates the procedure. If a selective solvent for the P4VP is used, cross-linking of the P4VP shell by, e.g., 1,4-dibromobutane allows the preparation of hollow spheres after removal of the hydrogen-bonded side chains. The latter was actually accomplished for a slightly different system consisting of P4VP and a random copolymer of styrene and a hydroxyl-containing monomer (PSOH) [124]. Chloroform is a good solvent for both, and when a dilute solution of PSOH in chloroform was added to a dilute solution of P4VP in nitromethane, a nonsol-

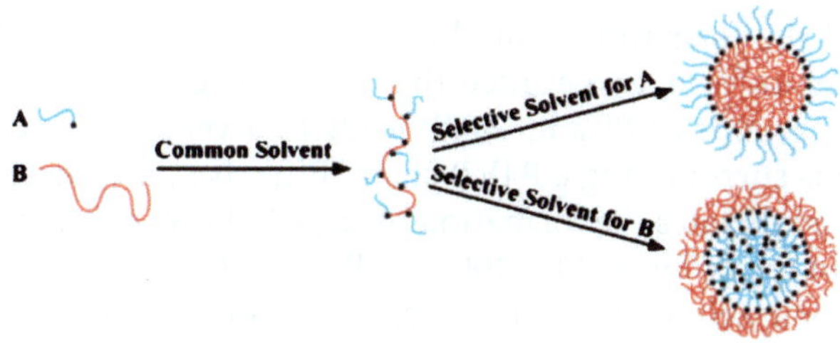

Fig. 13 Illustration of the formation of hydrogen-bonded comb copolymers and subsequent micellization in a selective solvent. Reprinted with permission from [127]. © 2005 American Chemical Society

vent for PSOH, micellization took place. After cross-linking of the P4VP shell, the PSOH could be removed by simply switching to the solvent DMF, which not only is a solvent for PSOH but also results in dissociation of the hydrogen bonds. This possibility to remove one of the hydrogen bonding species after it has served its purpose during the self-assembling process is an essential element that we will encounter over and over again.

Inspired by the results of Jenekhe and Chen [128], who demonstrated that rod–coil poly(phenylquinoline)-*block*-polystyrene diblock copolymers in a selective solvent for the coil-like polystyrene block self-assemble into hollow spherical micelles with a diameter of a few microns, Jiang and coworkers extended their strategy to rigid side chains [123, 127]. Rigid side chains will be discussed in more detail in the section dealing with hydrogen-bonded liquid-crystalline side-chain polymers, but here we will briefly review the self-assembly in dilute solution and the subsequent preparation of hollow spheres. Jiang and coworkers used two different rigid short chains, a carboxyl-terminated polyimide (PI) of $M_W = 4600$ and a cross-linkable carboxyl-terminated poly(amic acid) (PAE) of $M_W = 7450$ (see Scheme 5).

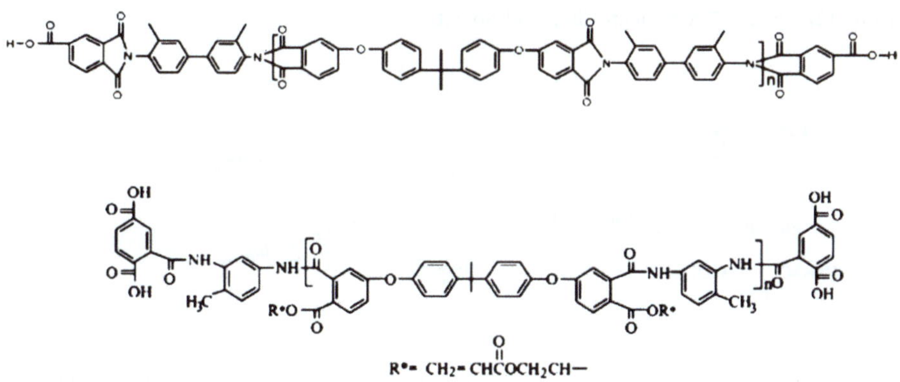

Scheme 5 Carboxyl-terminated polyimide and poly(amic acid) used in combination with P4VP to prepare hollow spheres [127]

They observed the formation of hollow spheres for P4VP and PI in a common solvent chloroform, and argued that the hydrogen bonding between the carboxylic end groups of PI and P4VP leads to a very high local concentration of PI chains surrounding a P4VP chain. Effective packing of the crowded rigid chains then dictates the formation of large hollow spheres with hydrodynamic radius in the order of 200–400 nm (P4VP: $M_w = 1.4 \times 10^5$ g/mol) [123, 127]. To fix such self-assembled structures photo-cross-linkable PAE was used [125]. Switching the medium from THF to THF/DMF (1 : 1) removed the P4VP shell, as witnessed by a significant decrease in size. Figure 14 illustrates both procedures.

In the next section we will extensively discuss block copolymer-based hydrogen-bonded side-chain polymers. Here we simply note that the above strategy to prepare hollow structures based on P4VP homopolymers can be easily extended to block copolymers. Jiang and coworkers used PS-*b*-P2VP di-

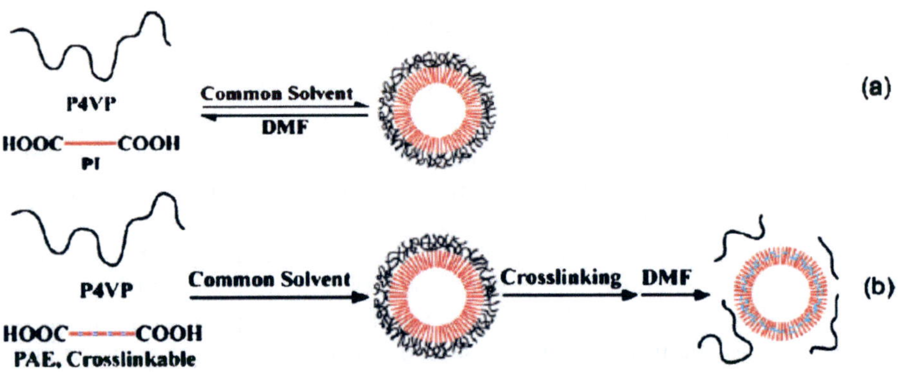

Fig. 14 Formation of hollow spheres by self-assembly of P4VP with PI and PAE, respectively. Subsequent cross-linking of PAE fixes the structure and the P4VP can be easily removed by adding DMF, thus breaking the hydrogen bonds. Reprinted with permission from [127]. © 2005 American Chemical Society

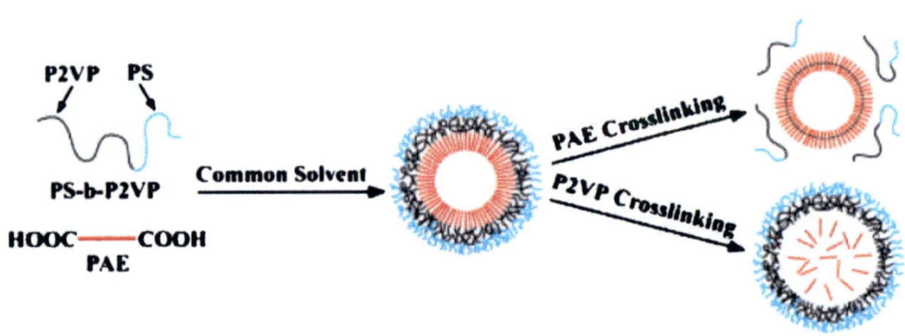

Fig. 15 Illustration of the preparation of hollow particles using PS-*b*-P2VP diblock copolymer together with PAE. Reprinted with permission from [127]. © 2005 American Chemical Society

block copolymers together with PAE to prepare hollow particles [126]. The procedure is similar to that discussed before and is illustrated in Fig. 15.

Micrometer-sized polymeric aggregates with three different morphologies (hollow spheres, solid spheres, and rods) were very recently reported from supramolecular assemblies of 5,7-dodecadiynedioic acid hydrogen bonded to P4VP [129]. Besides hydrogen bonding, the $\pi-\pi$ stacking of the diacetylenic moieties is argued to be responsible for the formation of these assemblies. The diacetylenic moieties can be polymerized by UV irradiation, resulting in blue-colored cross-linked assemblies that switch from blue to red upon heating.

4.2 Block Copolymer-Based Hydrogen-Bonded Comb Copolymers

4.2.1 Self-Assembly in the Bulk State: Two Length-Scale Structures

As already indicated, it is relatively straightforward to extend the procedure to form comb-shaped supramolecules via physical interactions, as discussed above for homopolymers, to diblock copolymers. The only requirement is that one of the blocks of the diblock copolymer allows physical bonding with suitable amphiphilic molecules. The best studied system consists of a diblock copolymer of poly(4-vinylpyridine) and polystyrene (P4VP-*b*-PS), where alkyl phenols such as pentadecylphenol (PDP) are hydrogen bonded to the P4VP block. Self-assembly now leads to hierarchically ordered structures characterized by two length scales: a large length scale due to microphase separation between the comb-shaped supramolecule block and the linear block, and a short length scale due to microphase separation within the domains formed by the comb-shaped supramolecule blocks [46, 130, 131]. In the given example the large length scale structure due to microphase separation between the linear PS block and the $P4VP(PDP)_{1.0}$ block exists up to temperatures of 200 °C and higher for molar masses of the PS and P4VP blocks exceeding 10 000 g/mol. The short length scale lamellar structure is formed on cooling around 65 °C in the P4VP(PDP)-containing domains. By selecting the relative block lengths of the P4VP-*b*-PS diblock copolymer appropriately, all the classical morphologies can be obtained thus leading to a series of structure-*within*-structure morphologies. The TEM micrographs presented in Fig. 16 prove this point for slightly larger nonadecylphenol (NDP) side chains.

All these morphologies have the disappearance of the short length scale lamellar ordering around 65 °C in common. On further heating, the fraction of hydrogen bonds gradually reduces. Furthermore, at ca. 130 °C PDP becomes completely miscible with PS. This implies that PDP, which in a sense is a P4VP-selective solvent at low temperatures, loses part of its selectivity. As a consequence, many of these systems show a series of order–order

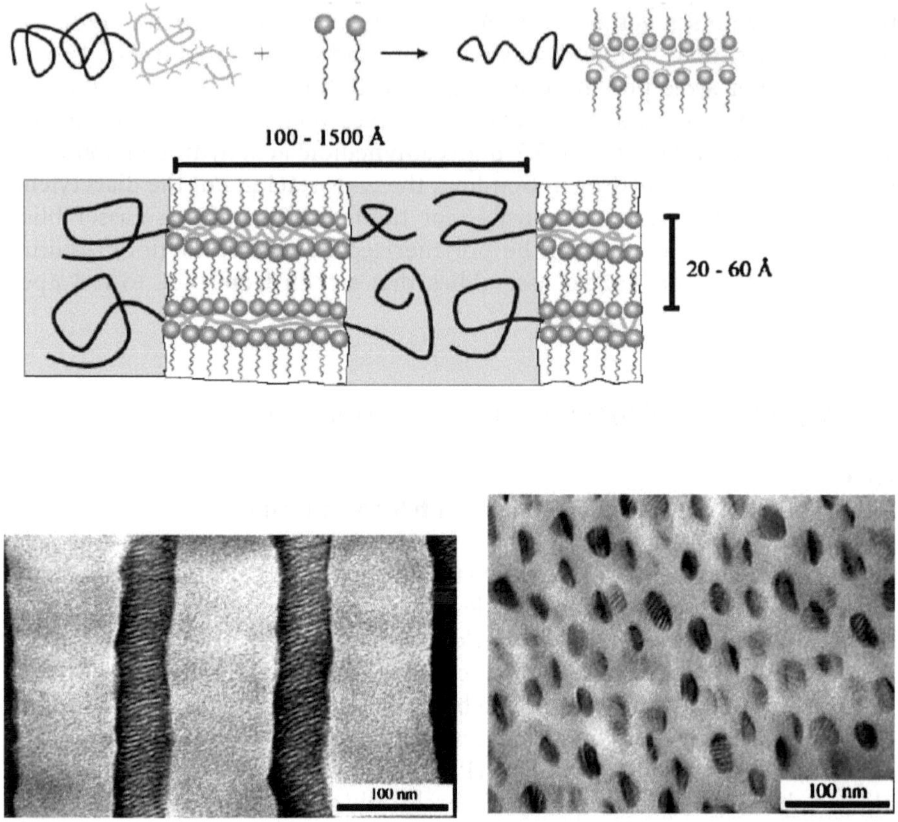

Fig. 16 Cartoon of lamellar-*in*-lamellar self-assembly and TEM pictures of lamellar-*in*-lamellar and lamellar-*in*-spheres morphologies for P4VP(NDP)$_{1.0}$-*b*-PS [131]

transitions for the large length scale structure. For instance, the system PS-*b*-P4VP(PDP)$_{1.0}$ with $M_{\mathrm{w,PS}} = 31\,900$ g/mol and $M_{\mathrm{w,P4VP}} = 13\,200$ g/mol, where the P4VP(PDP)$_{1.0}$ represents a 0.62 weight fraction, has a gyroid-*within*-lamellar structure at room temperature. At ca. 65 °C the lamellar structure of P4VP(PDP)$_{1.0}$ undergoes an order–disorder (ODT) transition and only the gyroid structure remains. At ca. 170 °C this transforms into a lamellar structure and finally becomes a hexagonally perforated layered structure at ca. 200 °C. Several other examples can be found in [132]. For PS-*b*-P4VP(PDP)$_{1.0}$ systems where the weight fraction of P4VP(PDP)$_{1.0}$ is in the range 0.50–0.60 and both blocks have a sufficiently large molar mass, the large length scale lamellar structure persists up to elevated temperatures. Now the diminishing degree of hydrogen bonding and the gradual diffusion of part of PDP into the PS layers manifests itself in the long period of the large length scale structure decreasing strongly during heating, in the order of 20% in the interval from 110 to 200 °C [133].

If the P4VP block of the P4VP-*b*-PS diblock copolymer is first turned into a polysalt by complexing with a strong acid such as MSA or TSA, as discussed before, hydrogen bonding with alkyl phenols such as PDP will give rise to a similar set of two length-scale structure-*in*-structure morphologies. Now the ODT of the short length scale structure is increased. However, especially in the case of MSA, there is an even stronger tendency for temperature-induced order–order transitions of the large length scale structure. Most interesting are the phenomena associated with the complex phase behavior in the P4VP homopolymer-based P4VP(MSA)(PDP) system discussed before. The macrophase separation between P4VP(MSA) and PDP, which occurs around 170 °C in the homopolymer-based system, is the driving force to push PDP from the P4VP(MSA)(PDP) domains into the PS domains in the diblock copolymer-based samples. Of course, the complete miscibility of PS and PDP above 130 °C is an essential element. Figure 17 presents temperature-dependent SAXS data of such a system, demonstrating a transition from a lamellar structure to a hexagonal structure. At elevated temperatures PDP is pushed into the PS layers, the volume fraction of these layers increases considerably at the expense of the P4VP(MSA)-containing domains, and the latter domains transform into cylinders [46]. Also clearly visible is the ODT of the short length scale structure around 100 °C, followed by a shift of the remaining correlation hole peak to smaller angles upon further heating (the q_1 peak). Figure 18 shows a characteristic series of TEM pictures for a different PS-*b*-P4VP$(MSA)_{1.0}(PDP)_{1.0}$ system undergo-

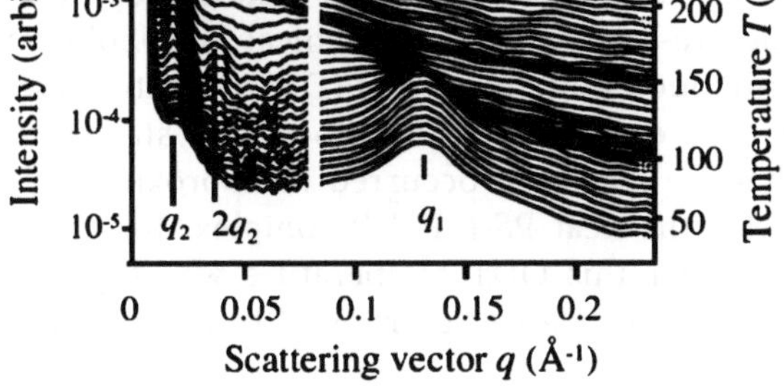

Fig. 17 SAXS of PS-*b*-P4VP$(MSA)_{1.0}(PDP)_{1.0}$ as a function of temperature. $M_{w,PS}$ = 40 000 g/mol, $M_{w,P4VP}$ = 5600 g/mol, weight fraction of P4VP$(MSA)_{1.0}(PDP)_{1.0}$ is 0.40. Clearly visible are the order–disorder transitions at ca. 100 °C and the order–order transition at ca. 150 °C [46]

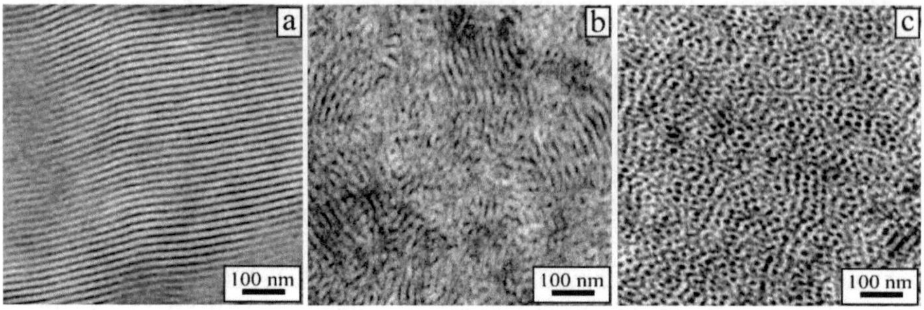

Fig. 18 TEM micrographs of PS-*b*-P4VP$(MSA)_{1.0}(PDP)_{1.0}$ system with $M_{w,PS} = 32\,900$ g/mol, $M_{w,P4VP} = 8100$ g/mol, weight fraction P4VP$(MSA)_{1.0}(PDP)_{1.0}$ is 0.54. **a** At room temperature a lamellar-*in*-lamellar structure, **b** at $T = 170\,°C$ P4VP cylinders, and **c** at $T = 210\,°C$ a spherical morphology. P4VP shows dark due to I_2 staining [132]

ing a series of large length scale lamellar to cylinders to spheres order–order transitions.

A few other examples of structure-*in*-structure morphologies have been reported. In one study [134] DBSA was complexed with a poly(1,4-butadiene)-*block*-poly(ethylene oxide) (PB-*b*-PEO; $M_{n,PB} = 11\,800$ g/mol, $M_{n,PEO} = 11\,000$ g/mol) diblock copolymer. For an average number of DBSA molecules of 0.5 and 1.0 per PEO monomer, a cylinder-*in*-lamellar morphology is obtained with PB cylinders of ca. 28 nm in diameter embedded in a lamellar self-assembled matrix with a ca. 2.9-nm-long period. The ODT of the PEO(DBSA) comblike blocks was found to occur at 45–55 °C and this transition is accompanied by an order–order transition from PB cylinders to PB spheres, i.e., a transition-driven transition. In subsequent papers Chen and coworkers [135, 136] investigated the effect of molecular architecture on the microphase separation of polystyrene-*block*-poly(2-vinylpyridine) with DBSA. A $(PS\text{-}b\text{-}P2VP)_5(PS)_5$ block-arm star copolymer and a $(PS)_5(P2VP)_5$ heteroarm star diblock copolymer complexed with DBSA were investigated and compared with a linear PS-*b*-P2VP diblock copolymer (Fig. 19). In both cases a structure-*in*-structure hierarchical morphology was found with PS cylindrical microdomains in a P2VP(DBSA) matrix exhibiting a short length scale lamellar ordering. In the block-arm star system the ODT of the short length scale structure occurred at approximately the same temperature as that for the neat PS-*b*-P2VP complex. In contrast, for the heteroarm star copolymer this ODT temperature was significantly increased, and this was attributed to a lower entropy of transition due to the junction constraint.

As a final example of a system exhibiting a typical two length-scale self-assembled structure we mention the acid–base complexes of poly(ethylene oxide)-*block*-poly(ethylene imine) (PEO-*b*-PEI) with dodecanoic acid investigated by Thünemann and coworkers [137]. A 2 : 1 ratio of amino func-

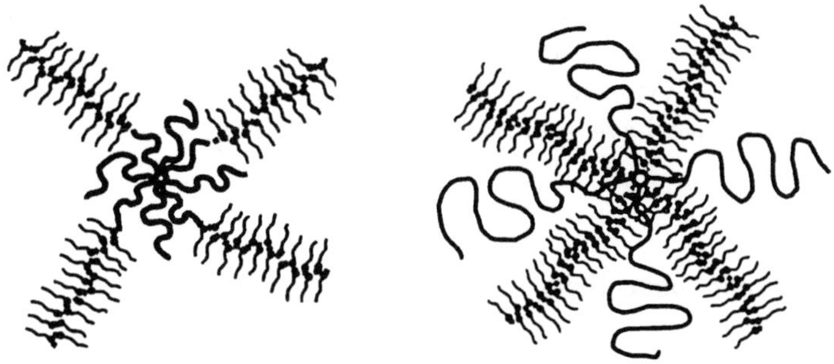

Fig. 19 Schematic representation of block-arm star copolymer of type (PS-*b*-P2VP)$_n$(PS)$_n$ and (PS)$_n$(P2VP)$_n$ heteroarm star diblock copolymer complexed with DBSA. In the drawing $n = 4$. Reprinted with permission from [135, 136]. © 2005 and 2006 American Chemical Society

tions to carboxylic acid functions was taken in order that approximately all carboxylic acid groups were attached to an amino group as confirmed by Fourier-transform infrared (FTIR) spectroscopy. The PEI blocks had different architectures (cyclic, linear, and branched) while the PEO blocks were kept the same. The block copolymers with linear and branched PEI self-assembled in characteristic lamellar-*in*-lamellar structures of, e.g., 15 and 3 nm, respectively.

4.3 Applications

4.3.1 Nanoscale Protonic Polymeric Conductors

Electroactive polymers range from conjugated polymers to ionically and protonically conducting materials. For ionic conductivity, solid polymer electrolytes involving lithium salts have aroused major interest, where suppression of the host polymer's crystallization (polyethylene oxide) increases the conductivity; block copolymers have been used as the suppressing agents [138, 139]. In solid protonic conductors, perfluorinated polymeric sulfonic acid materials, where the conductivity is due to proton hopping in adsorbed water, are most frequently used [140–143]. Another major type of protonically conductive polymers consists of acid–base complexes, typically salts of sulfonic acid-containing polymers blended with basic polymers [144, 145]. In protonic conductors, the potential of self-organization within the polymeric host has been discussed only recently. Above we discussed in some detail the comb-shaped polymeric supramolecules formed by hydrogen bonding alkyl phenols, such as pentadecyl phenol, to P4VP. When P4VP is complexed first with either MSA or TSA a proton-conducting acid–base complex is formed. The self-organization of these supramolecules gives rise to a lamellar structure. By covalently linking the P4VP chains to polystyrene blocks another level of self-assembly is introduced which, as we have seen, renders hierarchical structure-*in*-structure morphologies.

4.3.2 Switching Protonic Conductivity

As already discussed, interesting morphology transitions occur in PS-*b*-P4VP(MSA)$_{1.0}$(PDP)$_{1.0}$ samples as a function of temperature. One example of a series of different self-assembled structures, which corresponds to the SAXS pattern presented in Fig. 17, is schematically illustrated in Fig. 20. With respect to the conductivity, the main observation concerns the dimensionality of the P4VP(MSA)$_{1.0}$-containing proton-conducting domains. On increasing the temperature the one-dimensionally confined layers turn first into two-dimensional layers and then into one-dimensional cylinders. These transitions are directly reflected in the overall conductivity as demonstrated in the same figure [46, 146]. Self-assembly of block copolymer systems invariably leads to a multidomain grain boundary structure that is macroscopically isotropic. The conductivity presented therefore represents an average over all directions with respect to the local oriented structure.

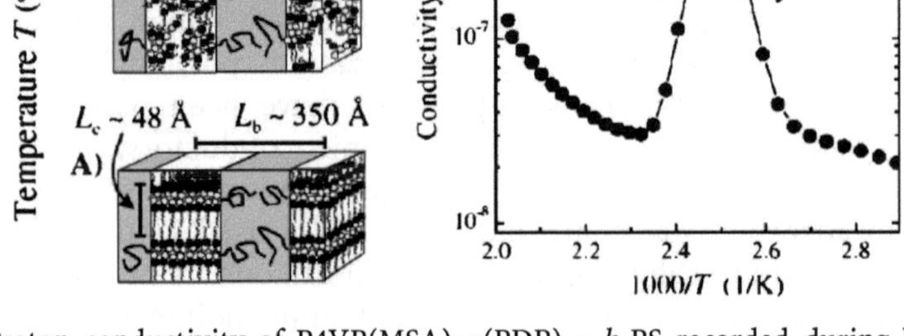

Fig. 20 Proton conductivity of P4VP(MSA)$_{1.0}$(PDP)$_{1.0}$-*b*-PS recorded during heating at 5 °C/min based on AC impedance measurements extrapolated to zero frequency. The cartoons show the dimensionality transitions occurring upon heating: from one-dimensional slabs to two-dimensional lamellae to one-dimensional cylinders. The order–disorder transitions at ca. 100 °C and the order–order transition at ca. 150 °C observed by SAXS (Fig. 17) are distinctly reflected in the conductivity [46]

4.3.3 Tridirectional Protonic Conductivity

Due to the macroscopic isotropic state of the sample just considered the conductivity is isotropic as well. In order to use the strong anisotropy of the local domains, the system has to be macroscopically oriented. This was achieved using oscillatory shear flow [147, 148], well known from studies on macroscopic orientation in conventional block copolymers [149–154]. Due to its thermal stability TSA was selected as the sulfonic acid to render $P4VP(TSA)_{1.0}$. To obtain the corresponding comb-shaped supramolecules, a stoichiometric amount of PDP versus sulfonate groups of $P4VP(TSA)_{1.0}$ was taken. The above P4VP was selected to be one block of a diblock copolymer PS-*b*-P4VP, which has an overall molar mass of $M_n = 50\,000$ g/mol, polydispersity 1.09, and weight fraction of PS $f_{PS} = 0.88$. Assuming that TSA and PDP are nominally fully complexed to form the supramolecules PS-*b*-$P4VP(TSA)_{1.0}(PDP)_{1.0}$, the weight fraction of the PS block is $f_{PS} = 0.62$. In this case self-assembly leads to alternating lamellae of PS and $P4VP(TSA)_{1.0}(PDP)_{1.0}$. Within the $P4VP(TSA)_{1.0}(PDP)_{1.0}$ layers there is an additional "inner" level of self-assembly, which consists of polar lamellae containing $P4VP(TSA)_{1.0}$ and nonpolar lamellae containing the pentadecyl chains of PDP. In contrast to the MSA case discussed above, the large length scale lamellar structure is stable up to temperatures as high as 200 °C. The ODT of the short length scale structure depends strongly on the amount of PDP used, and this made it possible to identify proper flow field conditions to achieve macroscopic orientation of both lamellar structures. Figure 21 presents the two-dimensional SAXS intensity patterns in tangential, normal, and radial directions demonstrating the level of macro-

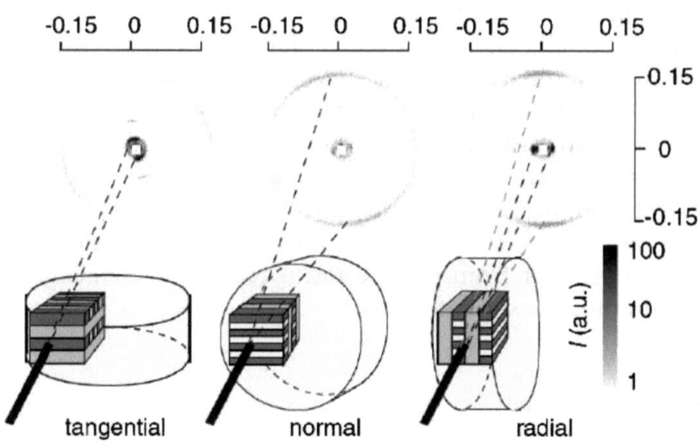

Fig. 21 SAXS of the self-assembled lamellar-*within*-lamellar structures of PS-*b*-P4VP $(TSA)_{1.0}(PDP)_{1.0}$ macroscopically oriented by oscillatory shear flow [155]

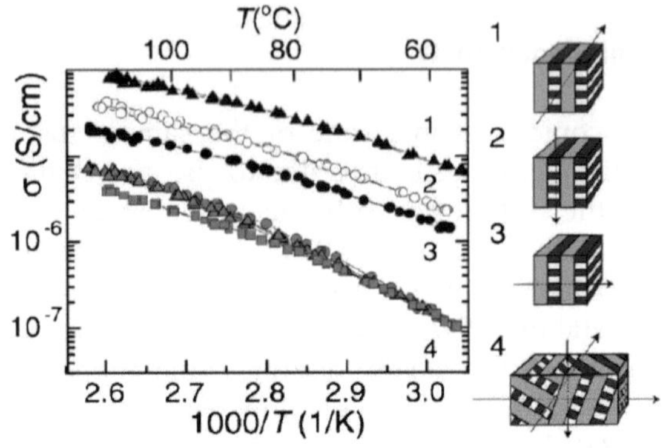

Fig. 22 Schematic self-assembled transverse-*within*-parallel aligned structure of PS-*b*-P4VP(TSA)$_{1.0}$(PDP)$_{1.0}$ supramolecules and the direct-current conductivity as a function of temperature. Presented also is the isotropic conductivity of a nonaligned sample (4). In the aligned case (3) the conductivity is increased, probably due to fewer domain boundaries, and the nanoscale conductivity anisotropy is manifestly present [155]

scopic orientation actually achieved [155]. Scattering peaks are observed at $q_1^* = 0.02\ \text{Å}^{-1}$ and $2q_1^*$, which demonstrates a lamellar self-assembly between PS and P4VP(TSA)$_{1.0}$(PDP)$_{1.0}$ with a periodicity of 32 nm. The lamellae are relatively well aligned along the shearing plates. SAXS shows also another peak at $q_2^* = 0.15\ \text{Å}^{-1}$ corresponding to the small lamellae with a long period of 4.1 nm. The SAXS picture demonstrates that the macroscopic orientation of the small length scale is transverse with respect to the flow field. For diblock copolymers this orientation sometimes occurs as a metastable state, the stable orientations corresponding to parallel or perpendicular depending on the flow field conditions [151, 152]. Here, however, we are dealing with a far more complex hierarchical lamellar-*in*-lamellar structure with a short length scale structure that aligns, at least under quiescent conditions, perpendicularly with respect to the large length scale structure [46]. The self-assembled structure with the corresponding tridirectional conductivity is presented in Fig. 22. However, the difference in conductivity in the three directions observed is rather small considering the materials involved. In particular, in the case of perfect alignment, one would expect to find hardly any conductivity in the direction perpendicular to the polystyrene layers. Clearly, the macroscopic orientation is far from perfect and it remains a challenge to achieve structures with much fewer defects.

4.3.4 Photonic Bandgap Materials

Self-assembly in block copolymer systems leads to well-ordered structures which are potentially interesting for photonic crystals applications. The transport of electromagnetic radiation can be manipulated using photonic bandgap materials, which contain periodic structures with sufficiently high dielectric contrast. In this case the transport of visible light may be totally or partially suppressed [156, 157]. There are, however, several important problems to be solved before this potential can be fully realized. One of these is the large periodicity required (order of $\lambda/4$, where λ is the wavelength of the electromagnetic radiation used). Several methods have been introduced to form photonic bandgaps, such as lithographic and etching techniques [158], colloidal self-assembly [159], synthetic opals [160–162], and inverted opals [163, 164]. Block copolymer self-assembly is another obvious possibility [165–168]. However, a large periodicity requires high molar mass block copolymers, which are notoriously difficult to prepare in a single crystal-like state. Here the comb-shaped supramolecules concept may have some advantages as well. Apart from inducing hierarchical structures in the case of block copolymers, the supramolecular side chains especially act as very efficient swelling agents. In a first example it was demonstrated that in the case of high molar mass PS-*b*-P4VP diblock copolymers, the PS-*b*-P4VP(DBSA) supramolecules form self-assembled one-dimensional optical reflectors [169]. The obvious next step consisted of exploiting the complex phase behavior of PS-*b*-P4VP(MSA)(PDP), discussed above in connection with switching proton conductivity, to prepare materials with reversible switching bandgaps [170]. Figure 23 shows how in this case the collapse of the periodic length scale at elevated temperatures of PS-*b*-P4VP$(MSA)_{1.0}(PDP)_{1.5}$

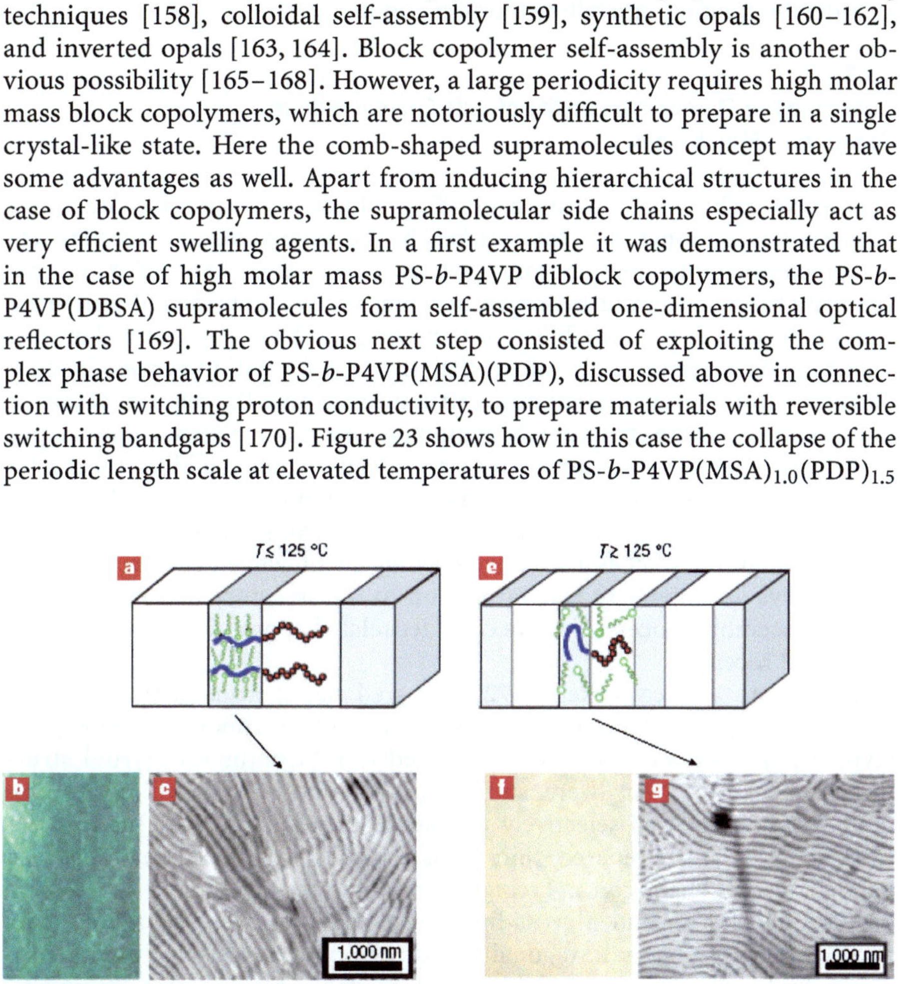

Fig. 23 PS-*b*-P4VP$(MSA)_{1.0}(PDP)_{1.5}$ at room temperature (**a–c**) and above 125 °C (**e–g**). At room temperature the large length scale periodicity is ca. 160 nm and the sample is *green*. Above ca. 125 °C, the sample becomes transparent due to a decrease in long period as a function of temperature, which results from the order–disorder transition in the P4VP$(MSA)_{1.0}(PDP)_{1.5}$ layers and a reduction in the number of hydrogen-bonded PDP side chains [171]

indeed results in reversible photonic-bandgap switching. This application also demonstrates how the very effective swelling by the side chains leads to the large periodicities required. Along a similar line Thomas and coworkers employed supramolecular side-chain liquid-crystalline (LC) block copolymers and showed that, upon heating the system above the LC isotropization temperature, the peak reflectivity changed by 40 nm resulting in a color change from green to orange [171]. The precise nature of this system will be discussed further on in the chapter in the section dealing with supramolecular side-chain liquid-crystalline block copolymers.

4.3.5 Exploiting the Thermoreversibility of Side Chain Bonding: Nanoporous Membranes

Thermoreversibility of hydrogen bonding allows for responsive functional properties as demonstrated, for example, by the temperature-induced switching of conductivity and photonic-bandgap switching. Another major advantage of the concept of hydrogen bonding-based comb-shaped polymeric supramolecules is the possibility of removing the hydrogen-bonded species after they have served their role in the structure formation process by simple dissolution. Because nanoporous structures require the removal of part of the material after the structure formation process is completed, comb-shaped diblock copolymeric supramolecules are ideally suited building blocks for functional organic polymeric membranes. The amphiphilic side chains can be removed in a straightforward way with a selective solvent. In this case the amphiphiles merely act as a selective swelling agent, while realizing, however, that it plays an important role in relation to the structure that is formed in the self-assembly process. First we consider related approaches that have been developed recently.

Hashimoto and coworkers [172] prepared nanoporous structures from a mixture of polystyrene-*block*-polyisoprene diblock copolymers and polystyrene homopolymers that self-assembled in a bicontinuous gyroid structure. The as-cast films were subjected to ozonolysis by which the isoprene component was selectively degraded. In a similar way, Liu and coworkers [173, 174] prepared thin films with densely hexagonally packed nanochannels. Their concept consists of diblock copolymers comprising a degradable block and a cross-linkable block. Block lengths are selected in such a way that a hexagonally ordered cylindrical structure is produced with cylinders formed by the degradable block. After microtoming the matrix is cross-linked and the cylinders are degraded. They prepared thin films with nanochannels from poly(*tert*-butyl acrylate)-*block*-poly(2-cinnamoylethyl methacrylate) (see Scheme 6), where the *tert*-butyl groups are cleavable by hydrolysis and poly(2-cinnamoylethyl methacrylate) is photo-cross-linkable. In a similar fashion Russell and coworkers [177]

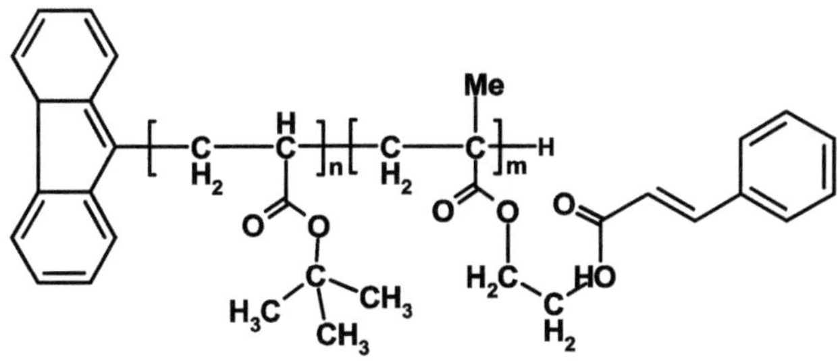

Scheme 6 Poly(*tert*-butyl acrylate)-*block*-poly(2-cinnamoylethyl methacrylate) [173]

prepared thin films of polystyrene-*block*-poly(methyl methacrylate) (PS-*b*-PMMA) self-assembled in hexagonally ordered PMMA cylinders of 14-nm diameter, and subsequently removed the PMMA component by deep ultraviolet exposure degrading the PMMA and at the same time cross-linking the PS matrix. In the case of cylinders, of major concern is their orientation with respect to the film surface. Usually a perpendicular orientation is aimed for. This can be accomplished by large amplitude oscillatory shear, as discussed before [148–154], which usually leads to a parallel to shear orientation, but which after microtoming produces thin films with perpendicularly aligned cylinders. Several alternative methods, which are more convenient for thin films, have been introduced during the last decade involving, e.g., electrical fields, special substrates, or controlled solvent evaporation. The group of Russell and coworkers has been particularly active in this area [175–182]. One example investigated resembles the hydrogen-bonded comb-shaped supramolecules discussed above. In hexagonally ordered thin films of diblock copolymers of poly(methyl methacrylate) (PMMA) and polystyrene (PS) (PS-*b*-PMMA) with PMMA cylinders, the cylinders were oriented perpendicular to the plane of the film using a specially modified substrate. By adding homopolymer PMMA of a slightly smaller molar mass than that of the PMMA block, the size of the cylindrical domains could be manipulated without perturbing the spatial order and orientation. Selective removal of the homopolymer produced pore diameters well below that achievable by pure block copolymers, whereas the removal of both the homopolymer and the corresponding block produced pores that are larger than achievable from the pure diblock copolymer [180].

The concept where amphiphilic molecules are hydrogen bonded to one block of a block copolymer to form specific comb-shaped supramolecules allows the production of nanoporous materials in a straightforward way. The concept is illustrated in Fig. 24. The procedure was first applied to a polystyrene-*block*-poly(4-vinylpyridine) diblock copolymer with a stoichiometric amount of pentadecylphenol hydrogen bonded to the P4VP block. The relative block lengths were selected to render a lamellar-*in*-cylindrical morphology, where the P4VP(PDP) comb-shaped blocks form the cylinders

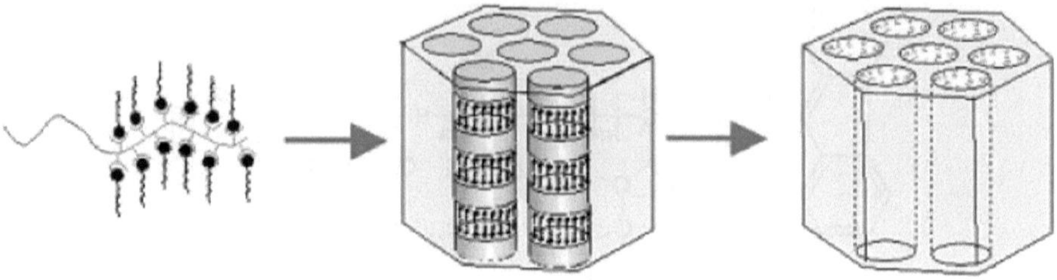

Fig. 24 Schematic illustration of nanoporous membrane preparation using the comb-shaped supramolecules concept [183]

within the rigid glassy PS medium and where the P4VP(PDP) complexes, being of comblike architecture, self-assemble as lamellae within the cylinders. When aiming for nanoporous membranes, however, the self-assembled structure inside the cylinders is not of interest. Perpendicular alignment of the cylinders was achieved by applying oscillatory shear followed by microtoming. Hollow channels with P4VP brushes at the interior walls were obtained in a straightforward manner by dissolving the PDP molecules away from the cylinders using methanol [183]. These "hairy tubes" open possibilities for controllable nanoporous membranes as the conformation of the brushes depends on the solvent. The diameter of the hollow channels can be tuned by selecting different block lengths of the diblock copolymer, but more importantly by adapting the amount of PDP added. In addition, the reactive P4VP chains at the hollow channel walls allow further chemical modification to produce nanofibers of, e.g., polypyrrole.

Subsequently, Stamm and coworkers [184–186] used PS-*b*-P4VP diblock copolymers in combination with 2-(4-hydroxybenzeneazo)benzoic acid (HABA, Scheme 7). Thin films with cylindrical nanodomains of the P4VP–HABA complexes in a PS matrix were produced. The alignment of the cylinders could be switched upon exposure to vapors of different solvents from parallel to perpendicular (Fig. 25). Extraction of HABA resulted in nanoporous membranes with hollow channels of 8 nm (Fig. 26). The channels were filled with Ni clusters via the electrodeposition method to fabricate an ordered array of metallic nanodots.

Besides hollow channels, porous lamellar structures have also been prepared via the comb-shaped supramolecules route. PS-*b*-P4VP diblock copolymers were taken that, together with zinc dodecylbenzenesulfonate [187],

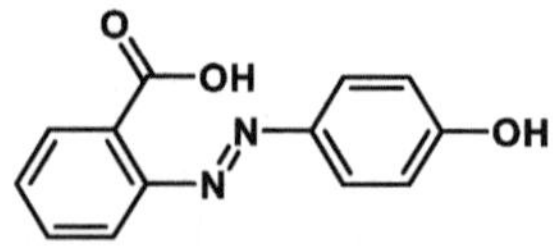

Scheme 7 2-(4-Hydroxybenzeneazo)benzoic acid (HABA) [185]

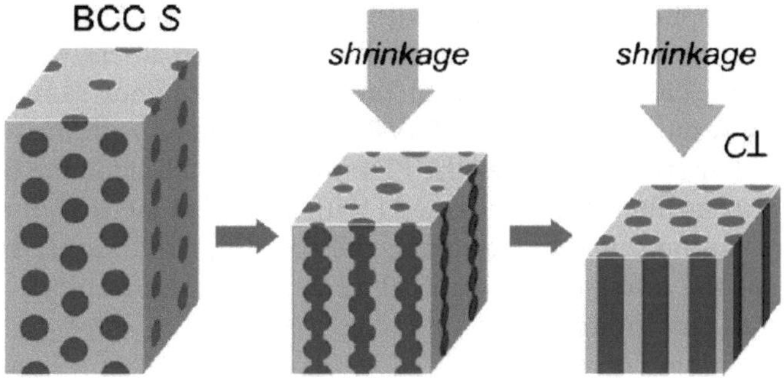

Fig. 25 Sketch of suggested transition from spherical morphology to perpendicularly oriented cylinders upon evaporation of the PS-selective solvent 1,4-dioxane. Reprinted with permission from [186]

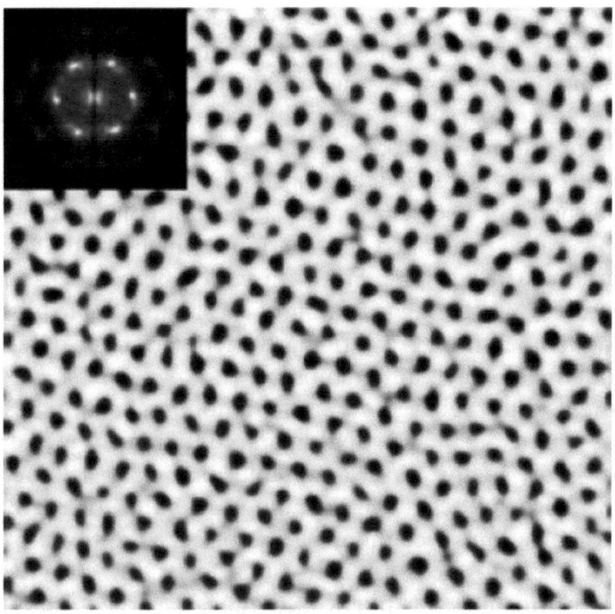

Fig. 26 AFM image with corresponding 2D fast Fourier transform plot of PS-*b*-P4VP(HABA) films after 1,4-dioxane vapor annealing to a swelling ratio of 3 and rinsing with methanol. Lateral scale of film $500 \times 500\,\text{nm}^2$ with a thickness of 61–68 nm. PS-*b*-P4VP: $M_n(\text{PS}) = 35.5 \times 10^3$ g/mol, $M_n(\text{P4VP}) = 3680$ g/mol, $M_w/M_n = 1.06$ for both blocks. Reprinted with permission from [186]

which forms a coordination complex with P4VP or with PDP [188], self-assemble in a large length scale lamellar structure. After extraction of the amphiphiles by a selective solvent the nanoporous structures are obtained. These were shown to be effective templates to direct the spatial organization of, e.g., metallic (Pd) nanoclusters [188].

In another application electrospinning [189] of PS-*b*-P4VP(PDP)$_{1.0}$ supramolecules was used to produce internally structured fibers with diameters in the range of 200–400 nm. Due to the block copolymer sample selected, self-assembly resulted in spherical P4VP(PDP) domains with the well-known internal lamellar structure. After the PDP was extracted from the fibers using methanol, porous fibers were obtained [190]. With this method, the thickness of the fibers can be tuned by adjusting the spinning conditions, and the size and nature of the pores can be controlled by the choice of block copolymer and amount of amphiphile.

4.3.6 Exploiting the Thermoreversibility of Side Chain Bonding: Nano-Objects

The nanoporous structures considered are frequently used as templates to produce, e.g., metallized nanowires or metallic films [177, 188]. On the other hand, self-assembly of block copolymers has been explored as well to prepare individual polymeric nano-objects. A general method is based on crew-cut aggregates, where amphiphilic diblock copolymers can be used, such as polystyrene-*block*-poly(acrylic acid) with a short hydrophilic block [82, 83, 191–194]. Crew-cut aggregates are typically constructed by first dissolving the chains in a solvent that dissolves both blocks and subsequently adding water to cause controlled aggregation of the hydrophobic blocks. Another option is to use a single solvent in which both blocks are soluble at high temperatures, and the aggregation of the hydrophobic block results on decreasing the temperature. A particularly rich variety of polymeric nanostructures is obtainable, including nanoscale rods.

Another method to prepare polymeric nanofibers, nanotubes, and spherical objects is based on block copolymers containing photo-cross-linkable moieties [195–197]. In this case, a suitable diblock or triblock copolymer is selected to allow the desired morphology in bulk, for example the hexagonal self-organization of cylinders if nanofibers are to be prepared. The concept uses the photo-cross-linkable block to fix the structure by photo-cross-linking. Ultimately, the fibers and tubes can be "sculptured" based on selective dissolution and/or degradation. Individual fibers of diameter ca. 40 nm were resolved and can even be redissolved in other solvents to allow liquid-crystalline solutions [196]. The advantage of this concept is that the design is based on the straightforward bulk phases of block copolymers instead of the more complicated solvent phases. However, there is a limitation that specific photo-cross-linkable moieties are required.

The comb-shaped supramolecules discussed at length in this review allow a novel and general concept to prepare crew-cut aggregates as well as more complex shaped disklike objects. The procedure to prepare crew-cut nano-objects is illustrated in Fig. 27 for the case of nanorods. The starting material is again poly(4-vinyl pyridine)-*block*-polystyrene. The corresponding comb–coil

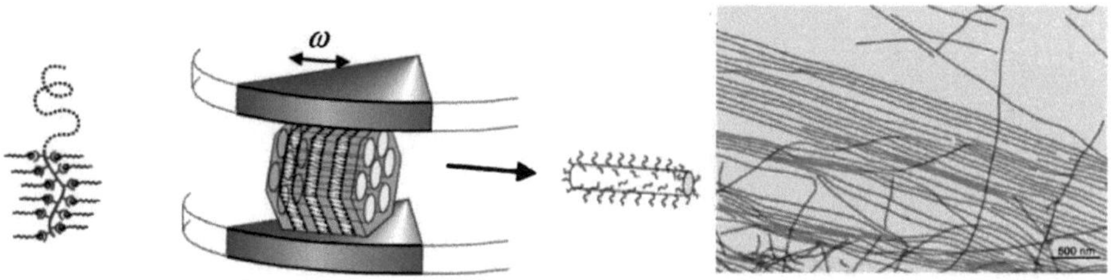

Fig. 27 Schematics showing the preparation of "hairy" nanorods [200]

supramolecules are obtained using P4VP-*b*-PS together with suitable hydrogen bonding amphiphiles such as the much-discussed PDP. It is relevant that the effective molar mass of the P4VP block can be increased in a controllable manner by complexation with side chains. This allows an easy way to tune the volume fraction of P4VP to achieve the desired morphology, e.g., PS cylinders. Because the side chains can be so easily removed by dissolution, they play a major role in the preparation of nano-objects in the present concept.

For the preparation of rodlike entities, only a cylindrical microdomain structure of the P4VP(PDP)-*b*-PS starting bulk material is required [198]. In order to obtain nanorods of several micrometers length, further improvement of the self-assembled structure by, e.g., large amplitude oscillatory shear is required [199, 200]. Figure 27 also shows a scanning transmission electron micrograph of the thus obtained nanorods with a PS core and a P4VP corona. The PS-*b*-P4VP nanorods were prepared from diblock copolymers of molar mass $M_{\mathrm{n,PS}} = 21\,400$ g/mol, $M_{\mathrm{n,P4VP}} = 20\,700$ g/mol, and $M_{\mathrm{w}}/M_{\mathrm{n}} = 1.13$ and have diameters of 25–28 nm with lengths up to several micrometers. They turned out to have rather poor mechanical properties. When a droplet of a nanorod suspension in ethanol was put on an alumina ultrafiltration membrane with characteristic pore size of 200 nm, those parts of the rods on top of the pores simply disappeared inside the pores due to capillary forces. These poor properties were attributed to the absence of entanglements. When a sufficient amount of homopolymer poly(2,6-dimethyl-1,4-diphenyl oxide) (PPE) with molar mass $M_{\mathrm{w}} = 25\,700$ g/mol was added, nanorods with a PS/PPE core and a P4VP corona were obtained with strongly improved mechanical properties [201]. PPE is well known for its excellent miscibility with PS [202] and because the molar mass between entanglements of PPE is only 4300 g/mol [203], entanglements are already introduced for weight fractions of 0.20 or higher.

The concept is not only useful to extract nanofibers, but also it can obviously be used for the preparation of other nano-objects, such as spheres and plates, when different effective values for the comb-block volume fraction are selected. In addition, the only specific requirement for the templating block copolymer is that one block has to be able to form strong physical bonds to allow formation of supramolecules. This allows one to "tune" the balance of

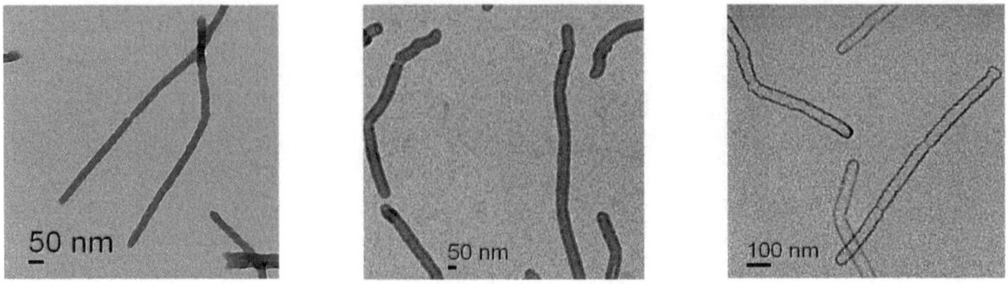

Fig. 28 *From left to right*: Scanning transmission electron micrograph of crew-cut nanorods consisting of PS core and P4VP corona nanorods, coated with aluminum oxide and after removal of the polymer material [206]

the composition of the block copolymer using additives that can easily be removed by solvent treatment. A practical point of view is that the additives should preferably have relatively high molecular weights to allow an efficient tuning of the effective composition.

As in the case of nanoporous films, nano-objects have been used as templates to prepare, e.g., hollow inorganic nanospheres and nanotubes. The concept has been used extensively by Wendorff and coworkers [204], who prepared metal and hybrid nano- and mesotubes by coating degradable polymer fibers obtained by electrospinning. For nanorods and nanospheres obtained via the hydrogen-bonded side-chain block copolymer route it has been introduced only very recently [205, 206]. The atomic layer deposition approach was used to template the polymeric nano-objects with a thin aluminum oxide layer, followed by removal of the polymer upon heating. Figure 28 shows TEM micrographs of the products obtained at the various steps of the process.

The foregoing procedure rests only on the thermoreversibility of the side chain bonding; the characteristic hierarchical structure-*in*-structure morphologies of the supramolecular comb–coil diblock copolymers played no role. Nevertheless, it is also possible to create nano-objects where, in addition to the thermoreversible side chain bonding, the two length-scale structures also play an essential role. As an example we mention the disklike objects prepared by Saito from lamellar-*in*-cylinders structures [207]. Here the starting material consisted of P4VP(PDP) cylinders in a PS matrix, where the cylinders consist in turn of alternating polar and nonpolar layers due to the comb-shaped P4VP(PDP) block. After cross-linking the P4VP disks with iodine and dissolving the PDP away, cross-linked P4VP disks with long PS hairs were obtained, objects that would be very difficult to construct otherwise.

4.4
Conjugated Polymer-Based Hydrogen-Bonded Comb Copolymers

Many of the most interesting polymers are intrinsically stiff, such as the electronically conducting conjugated polymers. These polymers typically include aromatic or heteroaromatic groups capable of π stacking, and generally do not melt and are poorly soluble in common solvents [208]. To improve processability flexible side chains have been attached to the rigid or wormlike polymer backbone, thus leading to so-called hairy-rod polymers [209–214]. The presence of side chains may result in characteristic liquid-crystalline behavior. Substituted poly(*p*-phenylene)s and poly(alkyl thiophene)s are examples where the hairy-rod structure leads to self-assembly with enhanced electronic properties [215–222].

In the case of rigid-rod polymers, it is in practical terms considerably more difficult to link side chains by hydrogen bonding. This is simply due to the fact that rigid polymers are highly incompatible with flexible chain molecules [223, 224]. As a consequence, the successful formation of a comb-shaped molecule, i.e., a hairy-rod supramolecule, usually requires stronger bonding. In the examples to be discussed, this is accomplished using a combination of ionic bonding and multiple hydrogen bonding.

In the limit of a rigid-rod backbone, the self-assembled states of hairy-rod molecules have been theoretically analyzed only very recently [225]. Figure 29 summarizes the different structures considered. From right to left the elastic stretching of the side chains accompanying microphase separation becomes increasingly more important due to an increase in the side chain length and/or side chain grafting density. As long as the stretching of the side chains is relatively unimportant a layered morphology should appear, whereas in the opposite case hexagonally ordered cylinders containing one or more hairy rods are predicted. The theoretical predictions have been put to the test in the recent work of Monkman and coworkers on the self-assembly of branched side chain hairy-rod polyfluorenes [226–228]. In the case of hairy-rod supramolecules the possibility of macrophase separation is another complicating factor. In fact, a theoretical analysis shows that such systems will be plagued by macrophase separation unless the composition is

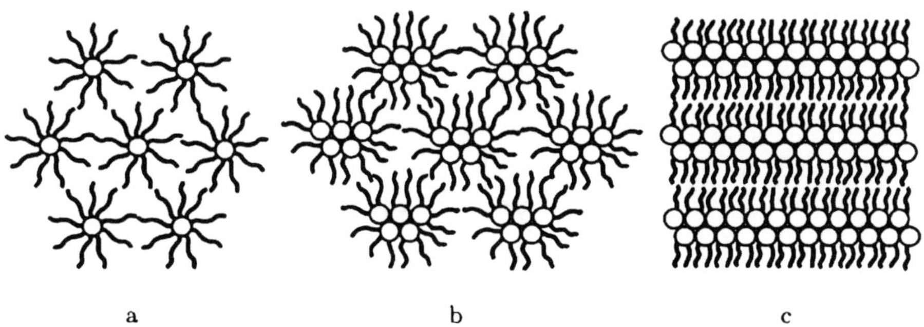

Fig. 29 Self-assembly of hairy-rod molecules. *From left to right* the interfacial tension becomes relatively more important compared to the elastic stretching of the side chains [225]

carefully chosen [229]. A relative excess of rod molecules will invariably lead to macrophase separation.

Besides the substituted poly(*p*-phenylene)s and the poly(alkyl thiophene)s, polyaniline (PANI) is one of the most interesting polymers in the area of electronically conducting polymers [230]. For PANI, protonation by strong acids is known to render conducting polymer salts. PANI is not really a rigid-rod polymer and it is expected that the electrical conductivity may greatly benefit from confinement of PANI chains within, e.g., narrow cylinders. This was achieved by the self-assembly of comb-shaped PANI supramolecules using a combination of ionic and hydrogen bonding [231]. This combination serves a dual purpose: the protonation introduces the electrical conductivity and the hydrogen bonding allows self-organization. The iminic nitrogens of PANI were first nominally fully doped using camphorsulfonic acid (CSA) to yield PANI(CSA)$_{0.5}$ and then hydrogen bonded to 4-hexylresorcinol (Hres) (Scheme 8). Both hydroxyl groups seem to be required to prevent macrophase separation. The resulting rodlike supramolecules self-assembled into hexagonally ordered cylindrical structures for a number y of Hres units per aniline repeat unit in the range $y = 0.5$–2.0. As demonstrated in Fig. 30, upon for-

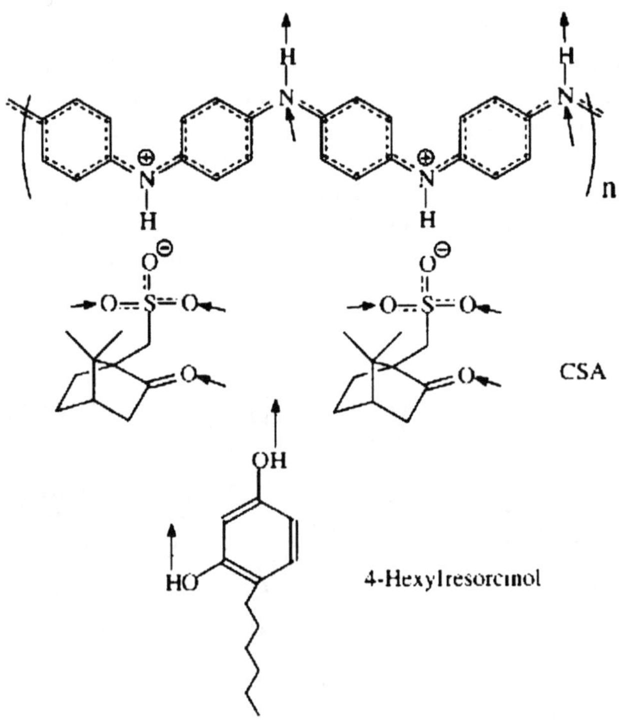

Scheme 8 Polyaniline (PANI), camphorsulfonic acid (CSA), and 4-hexylresorcinol (Hres) and proposed scheme of interaction [231]

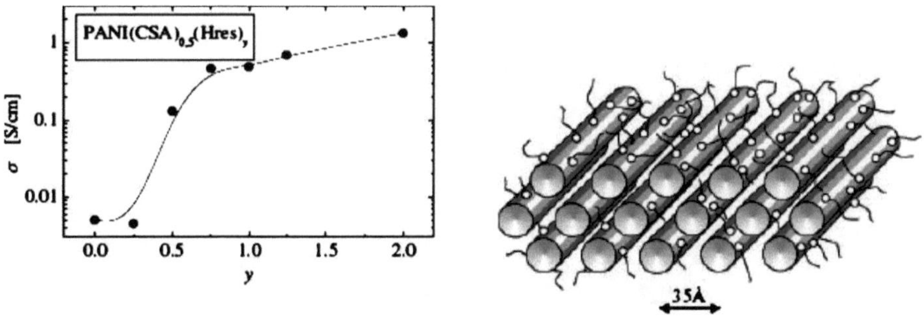

Fig. 30 Strongly enhanced electronic conductivity of $PANI(CSA)_{0.5}(Hres)_y$ after self-assembled cylinders are formed above a critical amount y of 4-hexylresorcinol [231]

mation of this structure, the electrical conductivity increased two orders of magnitude.

Another interesting example involves one of the simplest π-conjugated electroactive polymers poly(2,5-pyridine diyl) (PPY). The highly luminant conjugated polypyridine (PPY) is rodlike due to its *para*-coupled heteroaromatic rings. As such, it is soluble in only a few solvents, such as formic acid. After protonation by MSA the polymeric salt $PPY(MSA)_x$ is formed. Upon hydrogen bonding with selected alkyl phenols, such as octyl gallate (OG; octyl 3,4,5-trihydroxybenzoate), self-assembled nanostructures are formed (Scheme 9 and Fig. 31). This structure formation can be exploited to create thin solid films with interesting optical properties. For these kinds of systems the possibility to remove the hydrogen-bonded side chains in

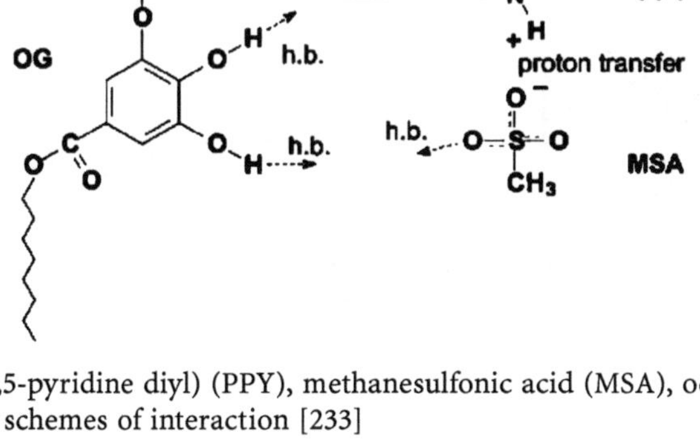

Scheme 9 Poly(2,5-pyridine diyl) (PPY), methanesulfonic acid (MSA), octyl gallate (OG), and the possible schemes of interaction [233]

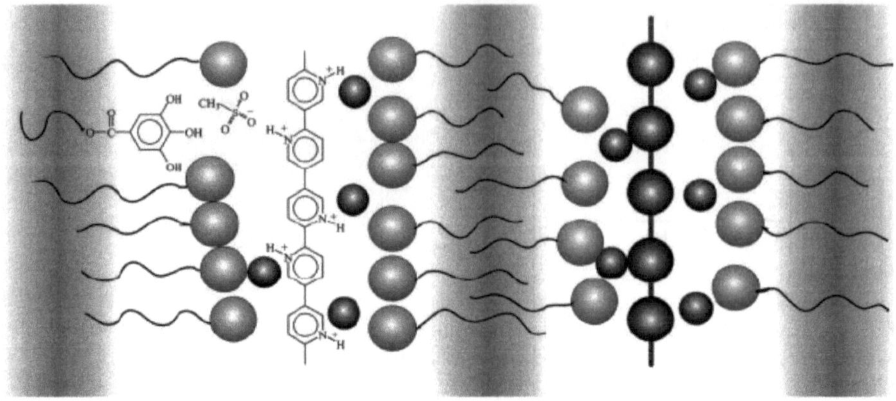

Fig. 31 Schematic lamellar structure comprising alternating polar layers (consisting of $PPY(MSA)_{1.0}$ and the aromatic parts of OG) and nonpolar layers formed by the octyl chains of OG [233]

a straightforward way offers yet another opportunity to create materials with specific function. To this end, every second PPY was complexed with CSA leading to $PPY(CSA)_{0.5}$. Limiting the mole fraction of CSA leads to relatively highly luminant complexes [232]. Suitable comb-shaped supramolecules were obtained by subsequent hydrogen bonding with Hres leading to $PPY(CSA)_{0.5}(Hres)_{0.5}$. As in the case of OG, this material self-assembles in the form of an exceptionally well-ordered layered structure [233–236]. The order can be classified as smectic liquid crystalline. The $PPY(CSA)_{0.5}$ layers can flow past each other and even a gentle sweep between two microscope glass plates suffices to achieve relatively high macroscopic orientation of the rods. Thereafter, the Hres molecules can be removed easily by mild heating in a vacuum oven, while still preserving the orientation of the rods. This brings the sample from the fluid self-assembled state to a solid film, which exhibits efficient polarized luminance with an efficiency of ca. 20%. Figure 32 illustrates this preparation scheme.

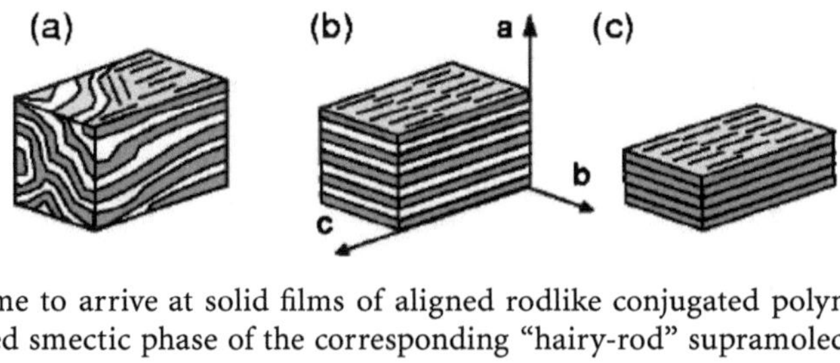

Fig. 32 Scheme to arrive at solid films of aligned rodlike conjugated polymers from the self-assembled smectic phase of the corresponding "hairy-rod" supramolecules [234]

5 Hydrogen-Bonded Comb Copolymers: Mesogenic Side Chains

Side-chain liquid-crystalline polymers combine polymeric properties with low molar mass mesogens. They are generally prepared by covalently linking mesogens via a flexible spacer group, which decouples at least partly the self-assembling tendency of the mesogens from the polymer conformation, to the polymer backbone [237–239]. In the last few decades alternative procedures have been introduced, in which the mesogens are linked to the polymer backbone by physical interactions, such as hydrogen-bonding, ionic, and charge-transfer interactions. [19–21]. Compared with covalently linked systems these supramolecular systems offer a much larger flexibility. Various parameters, such as the nature of the rigid core, the spacer length, and the nature and length of the terminal groups, can be varied relatively easily thus allowing fine-tuning of the liquid-crystalline properties. Carboxylic and benzoic acid groups are often used as H-bond donors together with pyridine or imidazole moieties.

5.1 Homopolymer-Based Hydrogen-Bonded Side-Chain Liquid-Crystalline Copolymers

Hydrogen-bonded side-chain liquid-crystalline polymers were first reported by Kato and Fréchet for binary mixtures of a polyacrylate polymer containing a 4-oxybenzoic acid unit linked to the backbone via a pentamethylene spacer group and a *trans*-stilbazole ester moiety with nitrogen at the *para* position [92]. Scheme 10a shows the extended mesogenic unit formed by simple single pyridine–benzoic acid hydrogen-bonded complexes. For all compositions nematic mesophases with higher transition temperatures than those of the individual components are observed. A similar procedure was used for the induction of liquid crystallinity in the side chain of polysiloxanes [240]. This is in fact a rather specific example involving the formation of an *extended* mesogenic structure. An alternative procedure, which resembles the P4VP(PDP) comb-shaped supramolecules discussed before, uses a mesogenic unit that is connected to an alkyl chain terminating in a functional group that functions as a hydrogen bonding molecular connector to the polymer backbone. An example is the hydrogen-bond donor poly(acrylic acid) (PAA) with cyanobiphenyl mesogens that are connected with an alkyl chain to imidazole-based hydrogen-bond acceptors (Scheme 10b) [241]. Self-assembly leads to smectic S_A structures. Network structures were obtained using the corresponding bis-imidazolyl compounds.

In a very recent publication Thomas and coworkers used the same principle with a custom-synthesized mesogen (LC) based on an imidazole head-group and a rigid biphenyl core with, however, a ten- and eight-carbon

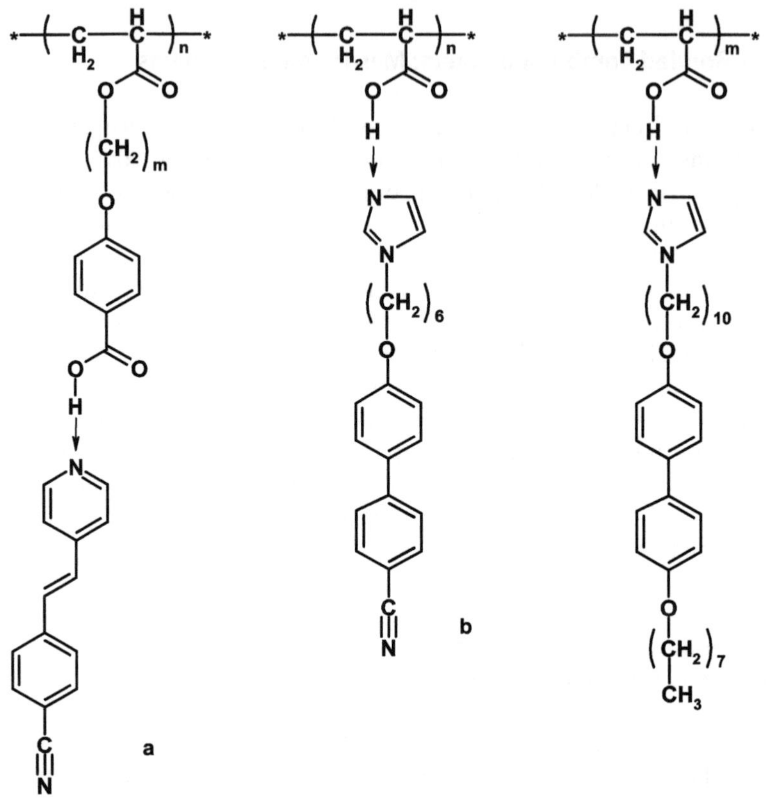

Scheme 10 Examples of hydrogen-bonded side-chain liquid-crystalline polymers [92, 241, 242]

aliphatic spacer and tail, respectively (Scheme 10c) [242]. A very rich phase behavior was observed for the substoichiometric 0.50 M PAA–LC. The small-angle X-ray data show unambiguously a large length scale microstructure within a small temperature window of 105–115 °C. Above 115 °C a liquid-crystalline structure of 75 Å is present due to the mesogens adopting a loosely packed bilayer arrangement. On cooling from the isotropic disordered phase at 150 °C, the 75-Å LC peak appears and first starts to recede at about 110 °C. At the same time a 120-Å peak appears together with its second-order peak. This new structure was formed reversibly on heating and cooling over a very small temperature range of 10–15 °C. The authors attributed it to the existence of differently complexed portions of the PAA backbone and the concurrent microphase separation between these portions. Figure 33 illustrates a homogeneous distribution to sequestering at one end of the PAA chain. A schematic picture of the molecular arrangement in the 105–115 °C range is given in Fig. 34. If true, this presents yet another aspect of microphase separation in hydrogen-bonded side-chain liquid-crystalline polymers.

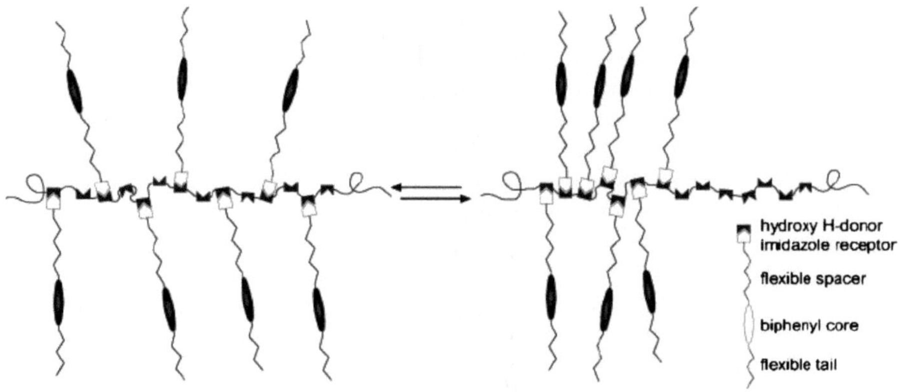

Fig. 33 Illustration of the two extremes, homogeneous and segregated, of the distribution of hydrogen-bonded side chains along the backbone in the case of a substoichiometric composition. System PAA(LC) investigated consists of poly(acrylic acid) and side chains containing cyanobiphenyl mesogens connected with an alkyl chain to imidazole-based hydrogen-bond acceptors (Scheme 10c). Reprinted with permission from [242]. © 2006 American Chemical Society

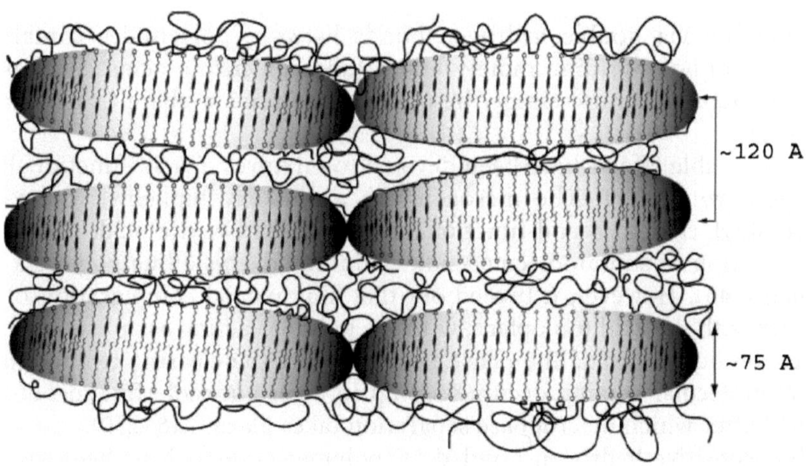

Fig. 34 Proposed microphase separated state of substoichiometric PAA(LC) with two characteristic length scales of 75 and 120 Å. Reprinted with permission from [242]. © 2006 American Chemical Society

Side-chain liquid-crystalline polymers were also prepared using double hydrogen bonds (Scheme 11) [243]. All these examples have in common a flexible spacer connecting the mesogenic unit to the polymer backbone. That this is not always required is demonstrated by yet another class of supramolecular liquid-crystalline polymers obtained by hydro-

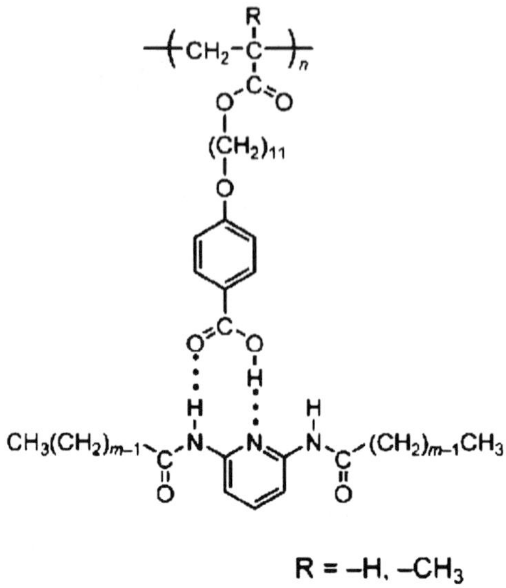

Scheme 11 Doubly hydrogen-bonded side-chain mesogens [19]

gen bonding via double hydrogen bonds between polyamides containing a 2,6-bis(amino) pyridine moiety and 4-(alkoxy)benzoic acids. Such polymeric structures cannot simply be classified as a main-chain or side-chain type [244].

Considerable efforts have been devoted to mixtures of acid-functionalized mesogens with P4VP. The mesogen was a dialkoxy biphenyl, where one of the alkyl chains is end-functionalized with carboxylic acid. Originally the formation of supramolecular side-chain liquid-crystalline polymers was claimed [245]. However, it turned out that only partial complexation of the polymer with the acid took place, with the polymer acting as an isotropic solute in the liquid-crystalline phase thus reducing the clearing temperatures. In fact molecular mixing occurs only up to a certain mole fraction of acid (ca. 0.3) after which macrophase separation takes place [246–250].

Photosensitive hydrogen-bonded LC polymer systems have been studied in some detail by Shibaev and coworkers [23, 251]. Nematic and smectic acrylic copolymers were used for the synthesis of hydrogen-bonded blends with photosensitive azobenzene-containing low molar mass species (Scheme 12). Blends of these copolymers with the photochromic dopants show nematic or smectic phase formation depending on the ratio of the components. The azobenzene-containing moieties undergo reversible *trans–cis* isomerization followed by orientation when exposed to plane-polarized laser irradiation. It was shown that, in principle, hydrogen-bonded photochromic blends may be used as optically active media for data recording [23, 252].

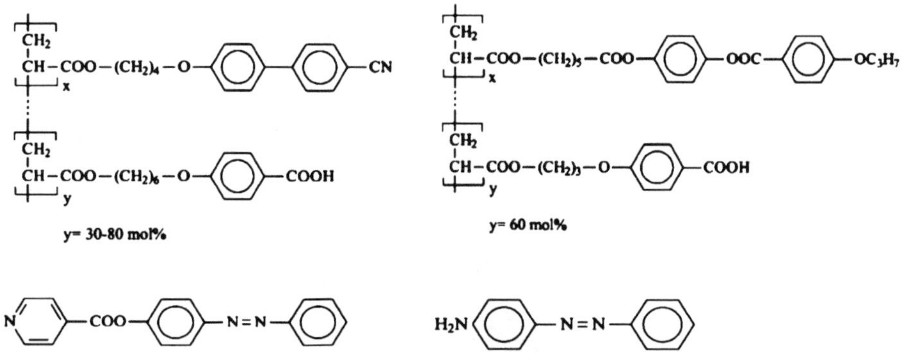

Scheme 12 *Top*: LC copolymers containing mesogenic side groups and carboxylic side groups capable of hydrogen bonding. *Bottom*: Azobenzene-containing dopants with pyridine ring and with a primary amine group ensuring the formation of donor–acceptor hydrogen bonds. Reprinted from [23, 252]. © 2003, with permission from Elsevier

5.2 Block Copolymer-Based Hydrogen-Bonded Side-Chain Liquid-Crystalline Copolymers

5.2.1 Temperature-Dependent Photonic Bandgap

As discussed before, the use of hydrogen-bonded side chains is a straightforward way to augment the periodicity of self-assembled block copolymers and bring it within the realm of photonic bandgap applications. In the case of mesogenic side chains leading to hydrogen-bonded liquid-crystalline block copolymers, there is the additional advantage that the photonic bangap can be tuned systematically, not only by the application of heat but also electric fields. Thomas and coworkers used a combination of polystyrene-*block*-poly(methacrylic acid) (PS-*b*-PMAA) diblock copolymers and imidazole-terminated mesogens (Scheme 13) [171]. They varied the amount of mesogen

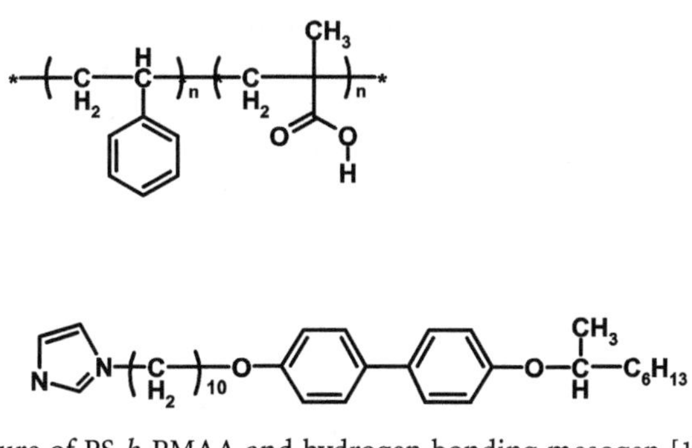

Scheme 13 Structure of PS-*b*-PMAA and hydrogen bonding mesogen [171]

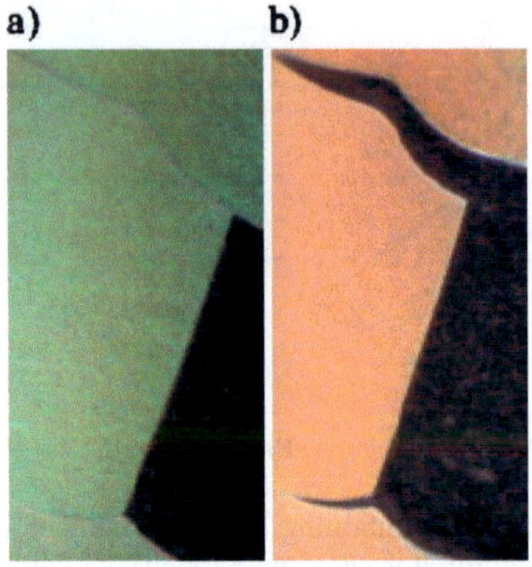

Fig. 35 Color change of polystyrene-*block*-poly(methacrylic acid) and imidazole-terminated side-chain mesogens PS-*b*-PMAA(LC) (Scheme 13) due to a shift in photonic bandgap accompanying isotropization [171]

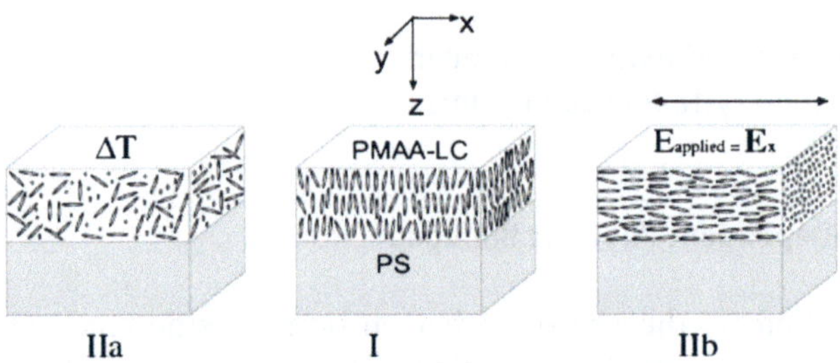

Fig. 36 Illustration of the effect of temperature on the original state I of PS-*b*-PMAA(LC). IIa: isotropization; IIb: electric field reorientation of the LC layers [171]

between 0.10 and 0.80 molar ratio of mesogen to acrylic acid repeat units. The neat PS-*b*-PMAA diblock copolymer forms hexagonally ordered cylinders of MAA in a PS matrix with an average cylinder–cylinder spacing of 120 nm. The PS-*b*-PMAA/LC material possessed a photonic bandgap in the green with the exact location depending on the composition. Heating the system above the LC isotropization temperature changed the peak reflectivity by 40 nm and resulted in a color change from green to orange (Fig. 35). In principle electric fields may also be used to dynamically manipulate the index contrast presented to the incident polarized light by switching the mesogen orientation (Fig. 36).

5.2.2
AC Orientational Switching

Thomas and coworkers demonstrated that orientational switching of mesogens and microdomains in certain hydrogen-bonded side-chain liquid-

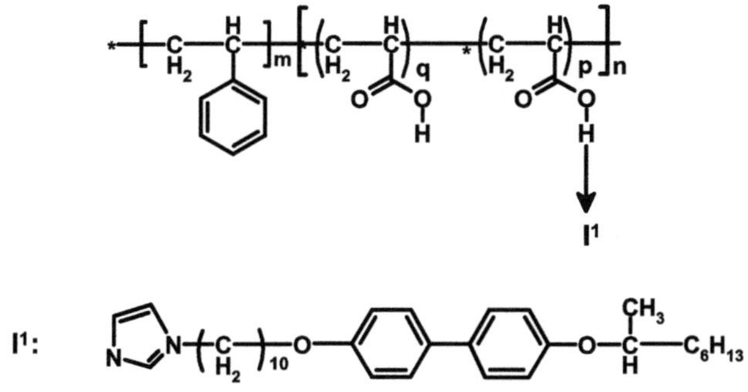

Scheme 14 Chemical structure of hydrogen-bonded side-chain liquid-crystalline block copolymer [28]

crystalline block copolymers can actually be achieved in minutes using moderate AC electric fields [28]. The fast orientation switching is argued to be directly related to the thermoreversibility of the hydrogen bonds. Diblock copolymers PS-*b*-PAA consisting of a polystyrene block of 25 000 g/mol and a poly(acrylic acid) block of 5300 g/mol were used and imidazole-terminated mesogens were attached to the PAA block by hydrogen bonding (Scheme 14). Two different compositions PS-*b*-PAA$(I1)_{0.5}$ and PS-*b*-PAA$(I1)_{0.8}$ were studied. Assuming all mesogens to be present in the PAA phase, these correspond to side-chain liquid-crystalline blocks of 35 000 and 24 000 g/mol, respectively. Both systems showed a lamellar microstructure due to a bilayer stacking of PS and PAA(I1). The ODT occurred at 180 °C for PS-*b*-PAA$(I1)_{0.5}$ and 170 °C for PS-*b*-PAA$(I1)_{0.8}$ with a 27-nm period for the latter. Furthermore, both samples showed smectic layers of 6 nm. The SAXS signature of these smectic mesophases disappeared at 110 °C. Interestingly, the glass transition temperature of the PS layers of PS-*b*-PAA$(I1)_{0.5}$ is at 85 °C, whereas it is around 70 °C for PS-*b*-PAA$(I1)_{0.8}$. The plasticization is attributed to the presence of mesogens in the PS phase with a higher amount for the high-mesogen-content blend. This is quite similar to the PS-*b*-P4VP$(PDP)_{1.0}$ systems discussed in the previous sections where the T_g of PS layers due to plasticization by PDP was found to be only 67 °C [253]. FTIR spectroscopy of PS-*b*-PAA(I1) demonstrated the presence of free mesogens. The number of free mesogens is higher for the high-mesogen-content blend and obviously increases as a function of temperature. The microstructure of both samples could be aligned in the perpendicular orientation when the sample was cooled down from a temperature above the ODT (170–180 °C) in an applied AC field. However, for PS-*b*-PAA$(I1)_{0.5}$ the orientation could not be changed from the starting parallel orientation induced by shear during the sample preparation to the perpendicular orientation at temperatures below the ODT, whereas this could be readily achieved for the PS-*b*-PAA$(I1)_{0.8}$ sample. As free mesogens readily align parallel to the applied AC field, this

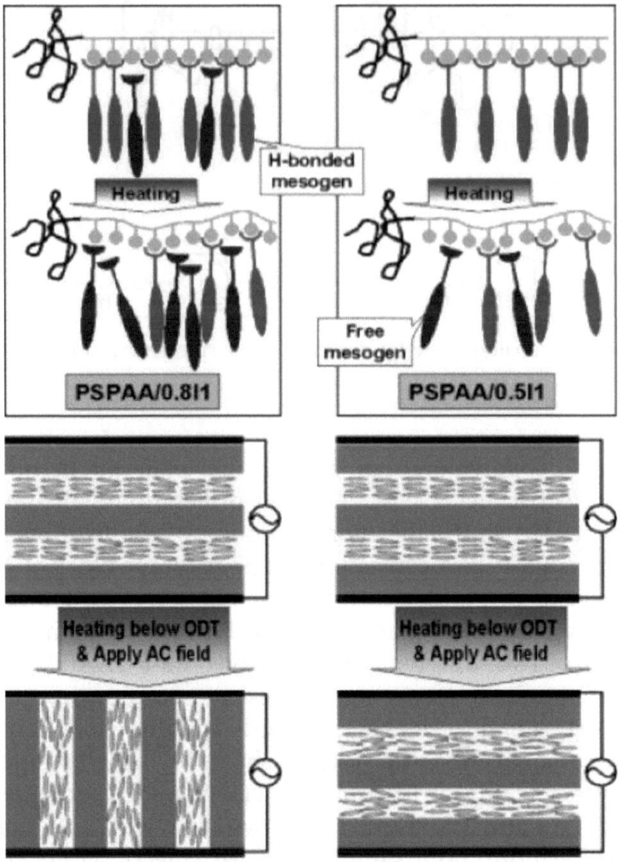

Fig. 37 Schematic illustration of the role of free mesogens of PS-*b*-PAA(LC) in the orientational switching by AC electric fields [28]

difference in behavior is attributed to the presence of a substantial number of free mesogens in the latter case (Fig. 37).

6 Layer-by-Layer Hydrogen Bonding Assembly

Layer-by-layer (LBL) deposition has become one of the major techniques to fabricate multilayer films. It was introduced by Decher in the form of alternate deposition of cationic and anionic polyelectrolytes [56–58]. This so-called electrostatic self-assembly process dominates the research in this area. However, more recently alternative driving forces have been employed, such as hydrogen bonding, coordination bonding, charge-transfer interactions, and covalent bonding. Since this review concerns the formation, functionalization, and tunability of ordered structures based on hydrogen bonding, we will focus on the former. LBL deposition based on hydrogen bonding was introduced almost simultaneously by Rubner et al. [51] and Zhang et al. [52, 254]. The latter reported the LBL approach using poly(4-vinylpyridine) and poly(acrylic acid) (Fig. 38). At about the same time, the Rubner group addressed the LBL adsorption of polyaniline with poly(vinylpyrrolidone) (PVPON), poly(vinyl alcohol) (PVA), poly(acrylamide), and poly(ethylene oxide). The resulting multilayer films consist of relatively thick, high polyaniline content bilayers due to the tendency of hydrogen bonding polymers to adsorb with a high density of loops and tails. This is in contrast to the electrostatic self-assembly process. Besides polymer pairs involving polyaniline as one of the components, polymer combinations with weaker hydrogen bonding capabilities such as PVPON and PVA were also investigated. Multilayer formation was not found to be possible and they concluded that self-assembly via hydrogen bonding requires polymers that are capable of forming strong hydrogen bonds. A potential advantage of the hydrogen bonding assembly is the possibility to use mixed solutions of, e.g., polyaniline and a hydrogen bonding polymer, in which case a single mixed layer is formed. In the case of electrostatic assembly this is obviously impossible due to the salt formation between the oppositely charged polymers and concurrent precipitation.

Hammond and DeLongchamp [255] reported the development of a highly ionic, conductive, solid polymer electrolyte film from hydrogen bonding LBL

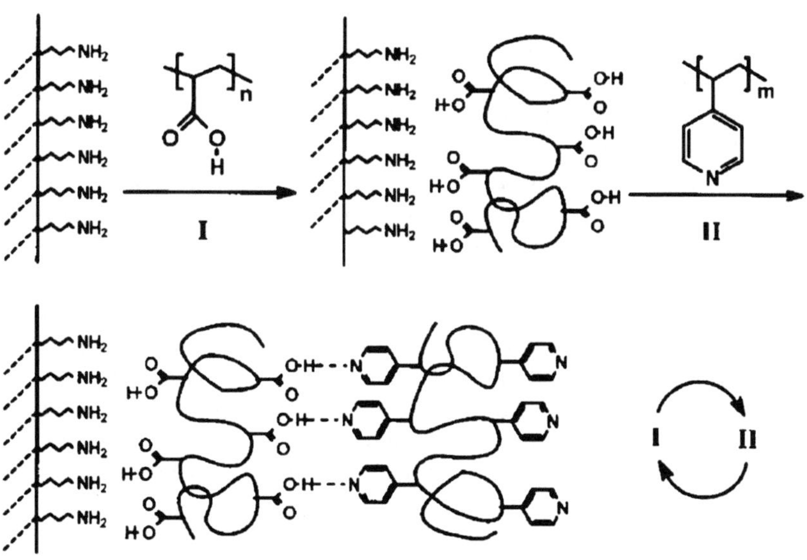

Fig. 38 Illustration of the LBL fabrication of multilayer films based on hydrogen bonding between PAA and P4VP. Reprinted with permission from [254]. © 1999 American Chemical Society

assembly of poly(ethylene oxide) and poly(acrylic acid). Films were self-assembled at a pH of 2.5 where PAA is fully protonated. Only around 10% of the carboxylic groups are estimated to participate in the hydrogen bonding with the ether oxygens. Because of this, low molar mass polymers have not sufficient multivalency for LBL assembly. The authors used PAA with $M_W = 90\,000$ g/mol and three different PEO samples of $M_W = 1500$, 20 000, and 4 000 000 g/mol. For the very low M_W, PEO films exhibited a roughness greater than the film thickness itself. For the higher molar mass PEO samples the average per-layer-pair thickness approached 100 nm at many deposition conditions, which is thicker than most electrostatically assembled LBL films. Arguments are presented that the films are almost homogeneous normal to the substrate other than a surface populated by the recently deposited film. The free ion population within PEO/PAA films could be greatly enhanced by exposure to lithium solutions. These hydrogen bonding-based LBL assembled polymers exhibit ionic conductivity exceeding that of electrostatically assembled LBL films investigated before, where ion pair sites create "Coulomb traps" slowing down free ion mobility [256]. They are potentially suitable for high-performance lithium polymer batteries.

The role of pH in the LBL assembly process based on hydrogen bonding has already been emphasized above. But also after the assembly, the multilayer film formed can be destroyed by environmental stimuli that introduce electrostatic charge into ionizable groups inside the film. Granick and coworkers demonstrated layered erasable films based on LBL deposition of PAA or PMAA, combined with PVPON or PEO [53, 54]. The multilayers turned out to be stable up to a system-specific pH and dissolved when the pH was raised above this point. Negative electric charges are introduced into the multilayers by ionization of the carboxylic acid groups, thus producing an electrostatic repulsion at ca. 14% ionization that could not be supported within an integral film. The transition turned out to be rather sharp.

The influence of a basic aqueous solution on LBL self-assembled films based on PAA and P4VP was investigated by Zhang and coworkers. They observed that in a first step the dissolution of PAA from the film into the solution occurred followed by a gradual conformational adaptation of the P4VP chains, thus producing a microporous film [25, 27]. In another study these authors examined the hydrogen-bonding-directed LBL assembly of P4VP and carboxyl-terminated polyether dendrimers [26]. Applying a basic solution resulted again in the formation of a microporous P4VP film, albeit of a different morphology, after the dissolution of the dendrimers.

The polymer pair PVPON and PMAA was later used by Sukhishvili and coworkers to prepare hydrogen-bonded multilayer capsules and to discuss the effect of the substrate charge on the first adsorbed layer of the weak poly(carboxylic acid) and the subsequent film growth [257]. Chen and coworkers used the same concept to prepare composite thin films by a hydrogen bonding assembly of polymer brushes and PVPON [258]. Spherical poly-

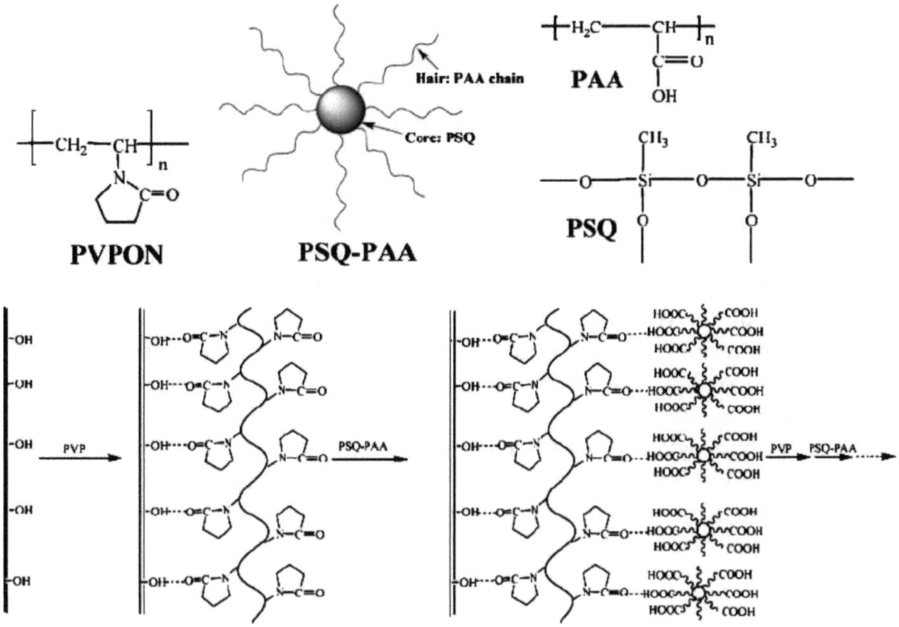

Fig. 39 Illustration of LBL fabrication of poly(vinylpyrrolidone) (PVPON) and spherical polymer brushes with a poly(methylsilsesquioxane) (PSQ) core and poly(acrylic acid) (PAA) corona. Reprinted with permission from [258]. © 2006 American Chemical Society

mer brushes with a poly(methylsilsesquioxane) (PSQ) core and PAA hairs were used sequentially as illustrated in Fig. 39. These films were subsequently claimed to get an inorganic film, by removing the organic material, or treated with tetrabutylammonium fluoride to remove the PSQ cores and get an organic film.

Lian and coworkers reported two routes to prepare Au nanoparticle multilayer thin films using two types of Au nanoparticles, respectively surface-modified with carboxylic and pyridine groups. In the first route carboxylic-functionalized Au particles and P4VP were deposited, whereas in the second route pyridine-functionalized Au particles and PAA were used. Similarly, CdSe nanoparticles with acrylic acid surface groups were deposited alternately with P4VP [259, 260].

7 Hydrogen-Bonded Interpenetrating Polymer Networks: Reversible Volume Transitions

The hydrogen bonding-based pseudo block copolymers considered were based on hydrogen bonding between end groups of different polymers to form a linear block copolymer architecture. Cooperative hydrogen bonding between complementary polymers has also been used to create responsive materials, notably polymer gels. Many gels undergo reversible discontinuous volume changes in response to environmental stimuli such as temperature, pH, etc. These transitions result from the competition between repulsive intermolecular forces that act to expand the polymer network and attractive forces opposing this. One of the best known examples involves attractive ionic interactions, such as those in a polyacrylamide gel with a pH-driven volume transition [261, 262]. In principle gels undergoing volume phase transition may be developed involving any of the four types of interaction: van der Waals, hydrophobic, ionic, and hydrogen bonding. The role of hydrogen bonding was first demonstrated by Tanaka and coworkers using interpenetrating polymer networks (IPNs) of poly(acrylic acid) and poly(acrylamide) [50]. These polymers interact via hydrogen bonding as illustrated in Scheme 15. Above a critical temperature the hydrogel swells due to a cooperative unzipping of the hydrogen bonds [262–264].

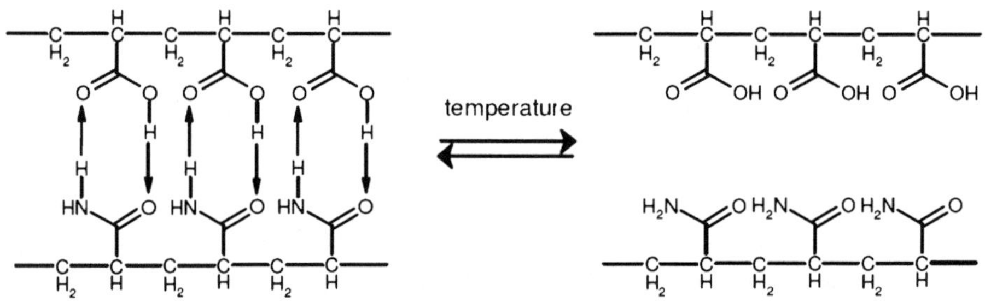

Scheme 15 Schematic picture of cooperative hydrogen bonding between poly(acrylic acid) (PAA) and poly(acrylamide) (PAAM) [264]

Stimuli-responsive polymeric hydrogel microspheres attract considerable attention due to their potential application in, e.g., controlled drug release. Most systems studied are based on poly(*N*-isopropylacrylamide) (PNIPAM), and exploit the LCST behavior of PNIPAM in water [265–271]. Hence, the particles deswell as a function of temperature. In a recent paper the synthesis of monodisperse thermoresponsive hydrogel microspheres was described consisting of a poly(acrylamide-*co*-styrene) core and poly(acrylamide)(PAAM)/PAA-based IPN shell with a positively thermoresponsive volume transition due to the cooperative hydrogen bonding between PAAM and PAA [264]. Figure 40 illustrates the corresponding temperature-induced volume change. Since this is likely to be the type of response looked for, the interest in hydrogen bonding-based hydrogels will only increase.

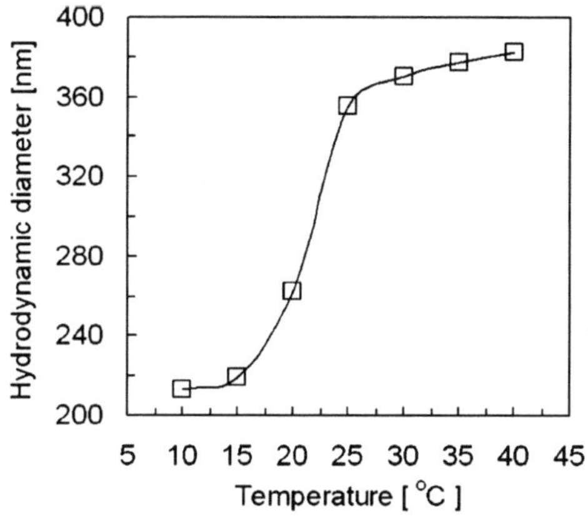

Fig. 40 Temperature-induced volume transition of core–shell microspheres with PAAM/PAA-based IPN shells. Reprinted from [264]. © 2005, with permission from Elsevier

8
Conclusion and Outlook

In supramolecular chemistry and in the area of polymer blends hydrogen-bonding interactions are among the most important noncovalent interactions. For polymer mixtures, hydrogen bonding allows for exothermic mixing thus compensating for the low entropy content of long-chain molecules. In supramolecular chemistry the strength, selectivity, and directionality play a dominating role. Both aspects, enhanced miscibility of chain molecules and the possibility to create hydrogen-bonded copolymers of specific macromolecular architectures, underlie the different subjects treated in this review. Hydrogen bonding-based miscibility is the essential ingredient for block copolymer blends, LBL deposition, and volume transitions in hydrogen-bonded IPNs. Strength, selectivity, and directionality are essential elements for the formation of hydrogen-bonded linear block copolymers as well as for the formation of hydrogen-bonded side-chain copolymers.

In virtually all the examples discussed multiple hydrogen bonding seems to be essential. For the hydrogen-bonded block copolymer blends, the hydrogen-bonded LBL deposition, and the hydrogen-bonded IPNs this is all too obvious. In the case of pseudo diblock copolymers a single hydrogen bond is certainly not strong enough to prevent macrophase separation between the two "long" incompatible telechelic macromolecules. In the case of the hydrogen-bonded side-chain polymer architecture a single hydrogen bond is sufficient to bind the side chain to the polymer; however, the polymer itself obviously has to contain many hydrogen bonding sites in order that a self-assembled periodic structure can be formed. In this review we highlighted the role of composition, temperature, and pH as essential parameters that influence the material properties. Composition is the essential parameter to direct the self-assembled morphology. Once the self-assembled state is present temperature is the most important external parameter for responsive properties, such as switching conductivity and switching photonic bandgaps. The role of pH has been discussed merely in relation to inducing charges in the self-assembled material, thereby reducing the number of hydrogen bonds and at the same time introducing repulsive forces. Examples where the hydrogen bond strength can be manipulated externally by means other than the temperature are scarce. Methods to modulate the strength of the hydrogen-bonding interaction between two given donor–acceptor groups have been discussed recently by Cooke and Rotello [22]. However, as concluded: "The majority of examples discussed involve the *synthetic* manipulation of the host unit's electrostatic properties and/or geometry to modulate the effectiveness of hydrogen-bonding interactions to a complementary guest." To create smart polymer materials based on hydrogen bonding the design of materials with electrochemical and photochemical control of the hydrogen bonding strength seems most interesting. In the present review a few examples were mentioned where electromagnetic fields were used to switch material properties, but it is expected that many more will be developed in the years to come.

References

1. Hamley IW (1998) The physics of block copolymers. Oxford Science Publications, Oxford
2. Fredrickson GH, Bates FS (1999) Phys Today 52:32
3. Muthukumar M, Ober CK, Thomas EL (1999) Science 277:1225
4. Fasolka MJ, Mayes AM (2001) Ann Rev Mater Res 31:323
5. Förster S, Plantenberg T (2002) Angew Chem Int Ed 41:688
6. Hamley IW (2003) Angew Chem Int Ed 42:1692
7. Liu T, Burger C, Chu B (2003) Prog Polym Sci 28:5
8. Park C, Yoon J, Thomas EL (2003) Polymer 44:6725
9. Abetz V, Simon PFW (2005) Adv Polym Sci 189:125
10. Hillmyer MA (2005) Adv Polym Sci 190:137
11. Lehn J-M (2005) Polym Int 51:825
12. Chao C-Y, Li X, Ober CK (2004) Pure Appl Chem 76:1337
13. Pollino JM, Weck M (2005) Chem Soc Rev 34:193
14. Binder WH (2005) Monatsh Chem 136:1
15. Ikkala O, ten Brinke G (2002) Science 295:2407
16. Liu S, Volkmer D, Kurth DG (2004) Pure Appl Chem 76:1847
17. Ikkala O, ten Brinke G (2004) Chem Commun 2131
18. ten Brinke G, Ikkala O (2004) Chem Rec 4:219
19. Kato T, Mizoshita N, Kishimoto K (2006) Angew Chem Int Ed 45:38
20. Kato T, Mizoshita N, Kanie K (2001) Macromol Rapid Commun 22:797
21. Kato T (2002) Science 295:2414
22. Cooke G, Rotello VM (2002) Chem Soc Rev 31:275
23. Shibaev V, Bobrovsky A, Boiko N (2003) Prog Polym Sci 28:729
24. Brunsfeld L, Folmer BJB, Meijer EW, Sijbesma RP (2001) Chem Rev 101:4071

25. Fu Y, Bai S, Cui S, Qiu D, Wang Z, Zhang X (2002) Macromolecules 35:9451
26. Zhang H, Fu Y, Wang D, Wang L, Wang Z, Zhang X (2003) Langmuir 19:8497
27. Bai S, Wang Z, Zhang X (2004) Langmuir 20:11828
28. Chao C-Y, Li X, Ober CK, Osuji C, Thomas EL (2004) Adv Mater 14:364
29. Leibler L (1980) Macromolecules 13:1602
30. De Gennes PG (1979) Faraday Discuss Chem Soc 68:96
31. Erukhimovich I (1982) Polym Sci Ser A 24:2223
32. Bates FS, Fredrickson GH (1990) Ann Rev Phys Chem 41:525
33. Binder K (1994) Adv Polym Sci 112:181
34. Matsen MW, Schick M (1994) Phys Rev Lett 72:2660
35. Matsen MW, Bates FS (1996) Macromolecules 29:1091
36. Hajduk DA, Harper PE, Gruner SM, Honeker CC, Kim G, Thomas EL, Fetters LJ (1994) Macromolecules 27:4063
37. Förster S, Khandpur AK, Zhao J, Bates FS, Hamley IW, Ryan AJ, Bras W (1994) Macromolecules 27:6922
38. Binder WH, Kunz MJ, Ingolic E (2003) J Polym Sci A Polym Chem 42:162
39. Kunz MJ, Hayn G, Saf R, Binder WH (2004) J Polym Sci A Polym Chem 42:661
40. Binder WH, Kunz MJ, Kluger C, Hayn G, Saf R (2004) Macromolecules 37:1749
41. Noro A, Nagata Y, Takano A, Matsushita Y (2006) Biomacromolecules 7:1696
42. Binder WH, Bernstorff S, Kluger C, Petraru L, Kunz MJ (2005) Adv Mater 17:2824
43. Takano A, Kawashima W, Noro A, Isono Y, Tanaka N, Dotera T, Matsushita Y (2005) J Polym Sci B Polym Phys 43:2427
44. Asari T, Matsuo S, Takano A, Matsushita Y (2005) Macromolecules 38:8811
45. Asari T, Arai S, Takano A, Matsushita Y (2006) Macromolecules 39:2232
46. Ruokolainen J, Mäkinen R, Torkkeli M, Mäkelä T, Serimaa R, ten Brinke G, Ikkala O (1998) Science 280:557
47. Coleman MM, Painter PC (1995) Prog Polym Sci 20:1
48. He Y, Zhu B, Inoue Y (2004) Prog Polym Sci 29:1021
49. Sperling LH (1994) In: Sperling LH, Klempner D, Utracki LA (eds) Interpenetrating polymer networks. Adv Chem Ser 239:3
50. Ilmain F, Tanaka T, Kokufuta E (1991) Nature 349:400
51. Stockton WB, Rubner MF (1997) Macromolecules 30:2717
52. Wang L, Wang Z, Zhang X, Shen J (1997) Macromol Rapid Commun 18:509
53. Sukhishvili SA, Granick S (2000) J Am Chem Soc 122:9550
54. Sukhishvili SA, Granick S (2002) Macromolecules 35:301
55. Tanaka T, Fillmore D, Sun S-T, Nishio I, Swislow G, Shah A (1980) Phys Rev Lett 45:1636
56. Decher G, Hong J-D (1991) Macromol Chem Macromol Symp 46:321
57. Decher G, Hong J-D, Scmitt J (1992) Thin Solid Films 210/211:831
58. Decher G (1997) Science 277:1232
59. Hashimoto T, Tanaka H, Hasegawa H (1990) Macromolecules 23:4378
60. Tanaka H, Hasegawa H, Hashimoto T (1991) Macromolecules 24:240
61. Winey KI, Thomas EL, Fetters LJ (1991) Macromolecules 24:6182
62. Shull KR, Mayes AM, Russell TP (1993) Macromolecules 26:3929
63. Semenov AN (1985) Sov Phys JETP 61:733
64. Mullens J, Yperman J, Francois JP, Van Poucke LC (1985) J Phys Chem 89:2937
65. Angerman HJ, ten Brinke G (1999) Macromolecules 32:6813
66. Shaknovich E, Gutin A (1989) J Phys (France) 50:1843
67. Panyukov S, Kuchanov S (1992) J Phys II (France) 2:1973
68. Fredrickson G, Milner S, Leibler L (1992) Macromolecules 25:6341

69. Erukhimovich I, Dobrynin A (1994) Macromol Symp 81:253
70. Walker JS, Vause CA (1987) Sci Am 256:98
71. Huh J, Jo WH (2004) Macromolecules 37:3037
72. Nor A, Nagata Y, Takano A, Matsushita Y (2006) Biomacromolecules 7:1696
73. Huh J, Park HJ, Kim KH, Kim KH, Park C, Jo WH (2006) Adv Mater 18:624
74. Folmer BJB, Sijbesma RP, Versteegen RM, Van der Rijt JAJ, Meijer EW (2000) Adv Mater 12:874
75. Hirschberg JHKK, Ramzi A, Sijbesma RP, Meijer EW (2003) Macromolecules 36:1429
76. Akiba I, Masunaga H, Sasaki K, Jeong Y, Sakurai K, Hara S, Yamamoto K (2004) Macromolecules 37:1152
77. Jiang S, Göpfert A, Abetz V (2003) Macromolecules 36:6171
78. Grunbaum B, Shepard GC (1986) Tilings and patterns. Freeman, New York
79. Tuzar Z, Kratochvil P (1993) In: Matijevic E (ed) Surface and colloid science, vol 15, chap 1. Plenum, New York, pp 1–83
80. Chu B (1995) Langmuir 11:414
81. Förster S, Zisenis M, Wenz E, Antonietti M (1996) J Chem Phys 104:9956
82. Zhang LF, Eisenberg A (1995) Science 268:1728
83. Zhang LF, Yu K, Eisenberg A (1996) Science 272:1777
84. Ding JF, Liu GJ, Yang ML (1997) Polymer 38:5497
85. Discher BM, Won Y-Y, Ege DS, Lee JCM, Bates FS, Discher DE, Hammer DA (1999) Science 284:1143
86. Pochan DJ, Chen Z, Cuii H, Hales K, Qi K, Wooley KL (2004) Science 306:94
87. Förster S (2005) Polymer vesicles. In: Encyclopedia of polymer science and technology. Wiley, New York
88. Kesselman E, Talmon Y, Bang J, Abbas S, Li Z, Lodge TP (2005) Macromolecules 38:6779
89. Hu J, Liu G (2005) Macromolecules 38:8058
90. Yan X, Liu G, Hu J, Willson CG (2006) Macromolecules 39:1906
91. Gao W-P, Bai Y, Chen E-Q, Li Z-C, Han B-Y, Yang W-T, Zhou Q-F (2006) Macromolecules 39:4894
92. Kato T, Fréchet JMJ (1989) Macromolecules 22:3818.
93. Ruokolainen J, Tanner J, ten Brinke G, Ikkala O, Torkkeli M, Serimaa R (1995) Macromolecules 28:7779
94. Valkama S, Lehtonen O, Lappalainen K, Kosonen H, Castro P, Repo T, Torkkeli M, Serimaa R, ten Brinke G, Leskelä M, Ikkala O (2003) Macromol Rapid Commun 24:556
95. Antonietti M, Conrad J, Thünemann A (1994) Macromolecules 27:6007
96. Antonietti M, Conrad J (1994) Angew Chem Int Ed Engl 33:1869
97. Antonietti M, Burger C, Effing J (1995) Adv Mater 7:751
98. Ikkala O, Ruokolainen J, ten Brinke G (1995) Macromolecules 28:7088
99. Ponomarenko EA, Waddon AJ, Tirrell DA, MacKnight WJ (1996) Langmuir 12:2169
100. Tsiourvas D, Paleos CM, Skoulios A (1997) Macromolecules 30:7191
101. Ober CK, Wegner G (1997) Adv Mater 1:17
102. Zhou S, Chu M (2000) Adv Mater 12:545
103. Thünemann AF (2002) Prog Polym Sci 27:1473
104. Antonietti M, Wenzel A, Thünemann A (1996) Langmuir 12:2111
105. Antonietti M, Maskos M (1996) Macromolecules 29:4199
106. Ruokolainen J, ten Brinke G, Ikkala O, Torkkeli M, Serimaa R (1996) Macromolecules 29:3409
107. ten Brinke G, Ruokolainen J, Ikkala O (1996) Europhys Lett 35:91

108. Ruokolainen J, Torkkeli M, Serimaa R, Vahvaselkä S, Saariaho M, ten Brinke G, Ikkala O (1996) Macromolecules 29:6621
109. Ruokolainen J, Torkkeli M, Serimaa R, Komanschek BE, Ikkala O, ten Brinke G (1996) Phys Rev E 54:6646
110. Ruokolainen J, Torkkeli M, Serimaa R, Komanschek BE, ten Brinke G, Ikkala O (1997) Macromolecules 30:2002
111. Eichhorn K-J, Fahmi A, Adam G, Stamm M (2003) J Mol Struct 661–662:161
112. Huh J, Ikkala O, ten Brinke G (1997) Macromolecules 30:1828
113. Luyten MC, Alberda van Ekenstein GOR, ten Brinke G, Ruokolainen J, Ikkala O, Torkkeli M, Serimaa R (1999) Macromolecules 32:4404
114. Ruokolainen J, Tanner J, Ikkala O, ten Brinke G, Thomas EL (1998) Macromolecules 31:3532
115. Ruokolainen J, Torkkeli M, Serimaa R, Vahvaselkä S, Saariaho M, ten Brinke G, Ikkala O (1996) Macromolecules 29:6621
116. Jiao H, Goh SH, Valiyaveettil S (2002) Langmuir 18:1368
117. Chen H-L, Ko C-C, Lin T-L (2002) Langmuir 18:5619
118. Ikkala O, Ruokolainen J, Torkkeli M, Tanner J, Serimaa R, ten Brinke G (1999) Colloids Surf A 147:241
119. Liu S, Zhang G, Jiang M (1999) Polymer 40:5449
120. Zhao H, Liu S, Jiang M, Yuan X, An Y, Liu L (2000) Polymer 41:2705
121. Liu S, Pan Q, Xie J, Jiang M (2000) Polymer 41:6919
122. Liu S, Jiang M, Liang H, Wu C (2000) Polymer 41:8697
123. Duan H, Chen D, Jiang M, Gan W, Li S, Wang M, Gong J (2001) J Am Chem Soc 123:12097
124. Wang M, Jiang M, Ning F, Chen D, Liu S, Duan H (2002) Macromolecules 35:5980
125. Kuang M, Duan H, Wang J, Chen D, Jiang M (2003) Chem Commun 496
126. Kuang M, Duan HW, Wang J, Jiang M (2004) J Phys Chem B 108:16023
127. Chen D, Jiang M (2005) Acc Chem Res 38:494
128. Jenekhe SA, Chen XL (1999) Science 283:372
129. Peng H, Lu Y (2006) Langmuir 22:5525
130. Ruokolainen J, Saariaho M, Ikkala O, ten Brinke G, Thomas EL, Torkkeli M, Serimaa R (1999) Macromolecules 32:1152
131. Ruokolainen J, ten Brinke G, Ikkala O (1999) Adv Mater 11:777
132. Valkama S, Ruotsolainen T, Nykänen A, Laiho A, Kosonen H, ten Brinke G, Ikkala O, Ruokolainen J (2006) Macromolecules 39:9327
133. Polushkin E, Alberda van Ekenstein GOR, Knaapila M, Ruokolainen J, Torkkeli M, Serimaa R, Bras W, Dolbnya I, Ikkala O, ten Brinke G (2001) Macromolecules 34:4917
134. Tsao C-S, Chen H-L (2004) Macromolecules 37:8984
135. Nandan B, Lee C-H, Chen H-L, Chen W-C (2005) Macromolecules 38:10117
136. Nandan B, Lee C-H, Chen H-L, Chen W-C (2006) Macromolecules 39:4460
137. Thünemann AF, General S (2001) Macromolecules 34:6978
138. Giles JRM, Gray FM, MacCallum JR, Vincent CA (1987) Polymer 28:1977
139. Lobitz P, Füllbier H, Reiche A, Illner JC, Reuter H, Höring S (1992) Solid State Ionics 58:41
140. Gray FM (1991) Polymer electrolytes. VCH, Weinheim
141. Kreuer KD (1996) Chem Mater 8:610
142. Agmon N (1995) Chem Phys Lett 244:456
143. Eikerling M, Kornyshev AA, Kuznetsov AM, Ulstrup J, Walbran S (2001) J Phys Chem B 105:3646
144. Kerres J, Ullrich A, Meier F, Häring T (1999) Solid State Ionics 125:243

145. Rikukawa M, Sanui K (2000) Prog Polym Sci 25:1463
146. Ikkala O, Ruokolainen J, Mäkinen R, Torkkeli M, Serimaa R, Mäkelä T, ten Brinke G (1999) Synth Met 102:1498
147. Mäkinen R, Ruokolainen J, Ikkala O, De Moel M, ten Brinke G, de Odorico W, Stamm M (2000) Macromolecules 33:3441
148. De Moel K, Mäkinen R, Stamm M, Ikkala O, ten Brinke G (2001) Macromolecules 34:2892
149. Keller A, Pedemonte E, Willmouth FM (1970) Colloid Polym Sci 238:25
150. Hadziioannou G, Mathis A, Skoulios A (1979) Colloid Polym Sci 257:136
151. Fredrickson GH, Bates FS (1996) Ann Rev Mater Sci 26:501
152. Chen Z-R, Kornfield JA, Smith SD, Grothaus JT, Satkowski MM (1997) Science 277:1248
153. Zhang Y, Wiesner U (1998) Macromol Chem Phys 199:1771
154. Chen Z-R, Kornfield JA (1998) Polymer 39:4679
155. Mäki-Ontto R, De Moel K, Polushkin E, Alberda van Ekenstein G, ten Brinke G, Ikkala O (2002) Adv Mater 14:357
156. Joannopoulos JD, Meade RD, Winn JN (1995) Photonic crystals. Princeton University Press, Princeton
157. Grier DG (ed) (1998) From dynamics to devices: directed self-assembly of colloidal materials. Mater Res Soc Bull (special issue) 23
158. Krauss TF, De La Rue RM (1999) Prog Quantum Electron 23:51
159. Lu Y, Yin Y, Xia Y (2001) Adv Mater 13:415.
160. Yoshino K, Kawagishi Y, Ozaki M, Kose A (1999) Jpn J Appl Phys 38:L788
161. Gates B, Park SH, Xia Y (2000) Adv Mater 12:653
162. Yoshino K, Satoh S, Shimoda Y, Kajii H, Tamura T, Kawagishi Y, Matsui T, Hidayat R, Fujii A, Ozaki M (2001) Synth Met 121:1459
163. Zakhidov AA, Baughman RH, Iqbal Z, Cui C, Khayyrullin I, Dantas O, Marti J, Ralchenko VG (1998) Science 282:897
164. Vlasov YA, Bo X-Z, Sturm JC, Norris DJ (2001) Nature 414:289
165. Fink Y, Urbas AM, Bawendy MG, Joanoppoulos JD, Thomas EL (1999) J Lightwave Technol 17:1963
166. Urbas A, Sharp R, Fink Y, Thomas EL, Xenidou M, Fetters LJ (2000) Adv Mater 12:812
167. Edrington AC, Urbas AM, DeRege P, Chen CX, Swager TM, Hadjichristidis N, Xenidou M, Fetters LJ, Joannopoulos JD, Fink Y, Thomas EL (2001) Adv Mater 13:421
168. Bockstaller M, Kolb R, Thomas EL (2001) Adv Mater 13:1783
169. Kosonen H, Valkama S, Ruokolainen J, Torkkeli M, Serimaa R, ten Brinke G, Ikkala O (2003) Eur Phys J E 10:69
170. Valkama S, Kosonen H, Ruokolainen J, Torkkeli M, Serimaa R, ten Brinke G, Ikkala O (2004) Nat Mater 3:872
171. Osuji C, Chao C-Y, Bita I, Ober CK, Thomas EL (2002) Adv Funct Mater 12:753
172. Hashimoto T, Tsutsumi K, Funaki Y (1997) Langmuir 13:6869
173. Liu G, Ding J (1998) Adv Mater 10:69
174. Liu G, Ding J, Hashimoto T, Kimishima K, Winnik FM, Nigam S (1999) Chem Mater 11:2233
175. Mansky P, Liu Y, Huang TP, Russell TP, Hawker CJ (1997) Science 275:1458
176. Huang E, Rockford L, Russell TP, Hawker CJ (1998) Nature 395:757
177. Thurn-Albrecht T, Schotter J, Kästle GA, Emley N, Shibauchi T, Krusin-Elbaum L, Guarini K, Black CT, Tuominen MT, Russell TP (2000) Science 290:2126
178. Jeong U, Kim H-C, Rodriguez RL, Tsai IY, Stafford CM, Kim KJ, Hawker CJ, Russell TP (2002) Adv Mater 14:274

179. Kim SH, Misner MJ, Xu T, Kimura M, Russell TP (2004) Adv Mater 16:226
180. Jeong U, Ryu DY, Kho DH, Kim JK, Goldbach JT, Kim DH, Russell TP (2004) Adv Mater 16:533
181. Kim SH, Misner MJ, Russell TP (2004) Adv Mater 16:2119
182. Wang J-Y, Xu T, Leiston-Belanger JM, Gupta S, Russell TP (2006) Phys Rev Lett 96:128301
183. Mäki-Ontto R, de Moel K, de Odorico W, Ruokolainen J, Stamm M, ten Brinke G, Ikkala O (2001) Adv Mater 13:107
184. Sidorenko A, Tokarev I, Minko S, Stamm M (2003) J Am Chem Soc 125:12211
185. Tokarev I, Krenek R, Burkov Y, Schmeisser D, Sidorenko A, Minko S, Stamm M (2005) Macromolecules 38:507
186. Tokarev I (2004) Order in thin films of diblock copolymers by supramolecular self-assembly. Thesis, Technical University Dresden
187. Valkama S, Ruotsalainen T, Kosonen H, Ruokolainen J, Torkkeli M, Serimaa R, ten Brinke G, Ikkala O (2003) Macromolecules 36:3986
188. Fahmi AW, Stamm M (2005) Langmuir 21:1062
189. Doshi J, Reneker DH (1995) J Electrostat 35:151
190. Ruotsolainen T, Turku J, Heikkilä P, Ruokolainen J, Nykänen A, Laitinen T, Torkkeli M, Serimaa R, ten Brinke G, Harlin A, Ikkala O (2005) Adv Mater 17:1048
191. Yu K, Zhang L, Eisenberg A (1996) Langmuir 12:5980
192. Desbaumes L, Eisenberg A (1999) Langmuir 15:36
193. Zhang L, Eisenberg A (1996) J Am Chem Soc 118:3168
194. Yu Y, Eisenberg A (1997) J Am Chem Soc 119:8383
195. Liu G (1997) Adv Mater 9:437
196. Liu G, Ding J, Qiao L, Guo A, Dymov BP, Gleeson JT, Hashimoto T, Saijo K (1999) Chem Eur J 5:2740
197. Stewart S, Liu G (2000) Angew Chem Int Ed Engl 39:340
198. De Moel K, Alberda van Ekenstein GOR, Nijland H, Polushkin E, ten Brinke G, Mäki-Ontto R, Ikkala O (2001) Chem Mater 13:4580
199. Alberda van Ekenstein G, Polushkin E, Nijland H, Ikkala O, ten Brinke G (2003) Macromolecules 36:3684
200. Fahmi AW, Brünig H, Weidisch R, Stamm M (2005) Macromol Mater Eng 290:136
201. Van Zoelen W, Alberda van Ekenstein G, Polushkin E, Ikkala O, ten Brinke G (2005) Soft Matter 1:280
202. Olabisi O, Robeson LM, Shaw MT (1979) Polymer–polymer miscibility. Academic, New York
203. Donald AM, Kramer EJ (1982) Polymer 23:1183
204. Bognitzki M, Hou H, Ishaque M, Frese T, Hellwig M, Schwarte C, Schaper A, Wendorff JH, Greiner A (2000) Adv Mater 12:637
205. Fahmi AW, Braun H-G, Stamm M (2003) Adv Mater 15:1201
206. Ras RHA, Kemell M, De Wit J, Ritala M, ten Brinke G, Leskelä M, Ikkala O (2006) Adv Mater (in press)
207. Saito R (2001) Macromolecules 34:4299
208. Skotheim TA, Elsenbaumer RL, Reynolds JR (1998) Handbook of conducting polymers. Dekker, New York
209. Ballauff M, Schmidt GF (1987) Makromol Chem Rapid Commun 8:93
210. Ballauff M (1989) Angew Chem Int Ed Engl 28:253
211. Wenzel M, Balluaff M, Wegner G (1987) Makromol Chem 188:2865
212. Vahlenkamp T, Wegner G (1994) Makromol Chem Phys 195:1933
213. McCarthy TF, Witteler H, Pakula T, Wegner G (1995) Macromolecules 28:8350

214. Lauter U, Meyer WH, Wegner G (1997) Macromolecules 30:2092
215. Guilleaume B, Blaul J, Ballauff M, Witteman M, Rehahn M, Goerigk G (2002) Eur Phys J E 8:299
216. Patel M, Rosenfeldt S, Ballauff M, Dingenouts N, Pontoni D, Narayanan T (2004) Phys Chem Chem Phys 6:2962
217. Gitsas A, Floudas G, Wegner G (2004) Phys Rev E 69:041802
218. Winokur MJ, Spiegel D, Kim Y, Hotta S, Heeger AJ (1989) Synth Met 28:C419
219. Samuelsen EJ, Mardalen J (1997) In: Nalwa HS (ed) Handbook of organic conductive molecules and polymers, vol 3. Wiley, New York, p 87
220. Breiby DW, Samuelsen EJ, Konovalov O, Struth B (2004) Langmuir 20:4116
221. Sirringhaus H, Brown PJ, Friend RH, Nielsen MM, Mechgaard K, Langeveld-Voss BMW, Spiering AJH, Janssen RAJ, Meijer EW, Herwig P, de Leeuw DM (1999) Nature 401:685
222. Sirringhaus H, Brown PJ, Friend RH, Nielsen MM, Mechgaard K, Langeveld-Voss BMW, Spiering AJH, Janssen RAJ, Meijer EW, Herwig P, de Leeuw DM (2000) Synth Met 111–112:129
223. Flory PJ (1978) Macromolecules 11:1138
224. Abe A, Ballauff M (1991) In: Ciferri A (ed) Liquid crystallinity in polymers. VCH, New York
225. Stepanyan R, Subbotin A, Knaapila M, Ikkala O, ten Brinke G (2003) Macromolecules 36:3758
226. Knaapila M, Kisko K, Lyons BP, Stepanyan R, Foreman JP, Seeck OH, Vainio U, Pälsson L-O, Serimaa R, Torkkeli M, Monkman AP (2004) J Phys Chem B 108:10711
227. Knaapila M, Stepanyan R, Lyons BP, Torkkeli M, Hase TPA, Serimaa R, Güntner R, Seeck OH, Scherf U, Monkman AP (2005) Macromolecules 38:2744
228. Knaapila M, Stepanyan R, Torkkeli M, Lyons BP, Ikonen TP, Almásy L, Foreman JP, Serimaa R, Güntner R, Scherf U, Monkman AP (2005) Phys Rev E 71:041802
229. Subbotin A, Stepanyan R, Knaapila M, Ikkala O, ten Brinke G (2003) Eur Phys J E 12:333
230. Chiang J-C, MacDiarmid AG (1986) Synth Met 13:193.
231. Koskonen H, Ruokolainen J, Knaapila M, Torkkeli M, Serimaa R, ten Brinke G, Bras W, Monkman A, Ikkala O (2000) Macromolecules 33:8671
232. Monkman AP, Halim M, Samuel IDW, Horsburgh LE (1998) J Chem Phys 109:10372
233. Ikkala O, Knaapila M, Ruokolainen J, Torkkeli M, Serimaa R, Jokela K, Horsburgh L, Monkman A, ten Brinke G (1999) Adv Mater 11:1206
234. Knaapila M, Ikkala O, Torkkeli M, Jokela K, Serimaa R, Dobnya IP, Bras W, ten Brinke G, Horsburgh LE, Palsson L-O, Monkman AP (2002) Appl Phys Lett 81:1489
235. Knaapila M, Torkkeli M, Jokela K, Kisko K, Horsburgh LE, Palsson L-O, Seeck OH, Dolbnya IP, Bras W, ten Brinke G, Monkman AP, Ikkala O, Serimaa R (2003) J Appl Crystallogr 36:702
236. Knaapila M, Stepanyan R, Horsburgh LE, Monkman AP, Serimaa R, Ikkala O, Subbotin A, Torkkeli M, ten Brinke G (2003) J Phys Chem B 107:14199
237. Ringsdorf H, Schneller A (1981) Br Polym J 13:430
238. Finkelmann H, Rehage G (1984) Adv Polym Sci 60/61:99
239. Shibaev VP, Plate NA (1984) Adv Polym Sci 60/61:173
240. Kumar U, Kato T, Fréchet JMJ (1992) J Am Chem Soc 114:6630
241. Kawakami T, Kato T (1998) Macromolecules 31:4475
242. Osuji CO, Chao C-Y, Ober CK, Thomas EL (2006) Macromolecules 39:3114
243. Kato T, Nakano M, Moteki T, Uryu T, Ujiie S (1995) Macromolecules 28:8875
244. Kato T, Ihata O, Ujiie S, Tokita M, Watanabe J (1998) Macromolecules 31:3551

245. Bazuin CG, Brandys FA (1992) Chem Mater 4:970
246. Brandys FA, Bazuin G (1996) Chem Mater 8:83
247. Stewart D, Imrie CT (1995) J Mater Chem 5:223
248. Alder KI, Stewart D, Imrie CT (1995) J Mater Chem 5:2225
249. Stewart D, Paterson BJ, Imrie CT (1996) Eur Polym J 33:285
250. Stewart D, Imrie CT (1997) Macromolecules 30:877
251. Barmatov E, Filippov A, Andreeva L, Barmatova M, Kremer F, Shibaev V (1999) Macromol Rapid Commun 20:521
252. Barmatov EB, Medvedev AV, Ivanov SA, Barmatova MV, Shibaev VP (2001) Polym Sci 43:285
253. Van Zoelen W, Alberda van Ekenstein G, Ikkala O, ten Brinke G (2006) Macromolecules 39:6574
254. Wang L, Fu Y, Wang Z, Fan Y, Zhang X (1999) Langmuir 15:1360
255. DeLongchamp DM, Hammond PT (2004) Langmuir 20:5403
256. DeLongchamp DM, Hammond PT (2003) Chem Mater 15:1165
257. Kozlovskaya V, Yakovlev S, Libera M, Sukhishvili SA (2005) Macromolecules 38:4828
258. Yang S, Zhang Y, Wang L, Hong S, Xu J, Chen Y (2006) Langmuir 22:338
259. Hao E, Lian T (2000) Langmuir 16:7879
260. Hao E, Lian T (2000) Chem Mater 12:3392
261. Tanaka T (1978) Phys Rev Lett 40:820
262. Tanaka T, Fillmore D, Sun S-T, Nishio I, Swislow G, Shah A (1980) Phys Rev Lett 45:1636
263. Endo N, Shirota H, Horie K (2001) Macromol Rapid Commun 22:593
264. Xiao X-C, Chu L-Y, Chen W-M, Zhu J-H (2005) Polymer 46:3199
265. Matsuoka H, Fujimoto K, Kawaguchi H (1999) Polym J 31:1139
266. Zhu PW, Napper DH (2000) Langmuir 16:8543
267. Pelton R (2000) Adv Colloid Interface Sci 85:1
268. Jones CD, Lyon LA (2000) Macromolecules 33:8301
269. Gan D, Lyon LA (2001) J Am Chem Soc 123:7511
270. Gao J, Hu Z (2002) Langmuir 18:1360
271. Zha L, Zhang Y, Yang W, Fu S (2002) Adv Mater 14:1090

Adv Polym Sci (2007) 207: 179–198
DOI 10.1007/12_2006_108

Published online: 9 December 2006

Nanocomposites Based on Hydrogen Bonds

Hao Xu · Sudhanshu Srivastava · Vincent M. Rotello (✉)

Department of Chemistry, University of Massachusetts, 710 N. Pleasant St., Amherst, MA 01003, USA
rotello@chem.umass.edu

1	**Introduction to Polymer/Nanocomposites**	179
2	**Principles of Nanocomposite Design**	181
2.1	Nanoparticles as Building Blocks	181
2.2	Polymer Scaffolds	182
2.3	Surface Modification and Patterning	183
3	**Control of Structure and Functionality**	184
3.1	Control of Interparticle Distance	185
3.2	Directed Assembly of Nanobuilding Blocks on Planar Substrates	187
3.3	Detailed Tailoring of Self-Organized Structures	190
3.4	Polymer and Nanoparticle 3D Aggregates in Solution	191
4	**Conclusion and Outlook**	195
References		196

Abstract Materials with nanoscale dimensions display electronic, photonic, and magnetic properties different from those observed by their respective bulk materials. This article describes the utilization of hydrogen bonds for modular self-assembly of nano-sized building blocks into two or three-dimensional aggregates and the precise control over their structural parameters and morphologies. We will depict recent investigations on the synthesis and assembly of polymer and nanoparticle-based composite materials both in solution phase and on solid substrates. The advantages, potential applications, and current challenges associated with this "bottom up" assembly approach will also be discussed.

Keywords Hydrogen-bonding · Self-assembly · Nanocomposites · Supramolecular chemistry · Bottom-up fabrication

1
Introduction to Polymer/Nanocomposites

Nanocomposites based on polymers and nanoparticles (NPs) are diverse and versatile functional materials, with applications ranging from electronic device fabrication to biosensors and catalysis [1–7]. The controllable polymer chain length, tunable NP size and core materials, and various side-chain and ligand functionalization enable these macromolecular building blocks to assemble into objects with various size, shape, composition, and surface structure [8]. In many instances, the ability to exploit the properties of nanocomposites for integrated devices and fabrication processes will require the formation of morphologically controlled and highly ordered arrays [9, 10]. The traditional "top-down" fabrication technique employs two-dimensional etching, deposition and layer-by-layer manufacturing processes that constitute the foundation of current microelectronic technologies. However, with the continuous requirement for high-performance and low-cost devices, this approach itself not only reaches its resolution limit due to light diffraction at nanometer scales, but is also inefficient in terms of energy and materials. Progress in our understanding of self-assembly process, which constitutes the "bottom-up" alternative approach has the potential of overcoming this technological barrier [11–21]. Self-assembly offers opportunities to simplify processes, lower costs, and develop new methods through the spontaneous generation of order in three dimensions (3D) or on curved surfaces [22].

One important focus of self-assembly is the tailoring of interfacial interactions [3] that ultimately determine the equilibrium geometry and spacing of the building blocks [8]. Noncovalent interactions are required to provide molecules enough mobility to self-assemble into ordered suprastructures without the need for chemical reactions to happen [23, 24]. Those interactions include hydrogen bonding, metal coordination, electrostatics, dipolar interactions, Van der Waals forces, and hydrophobic interactions. Hydrogen bonding plays an important role in the organization/reorganization of the biological structures, such as the hybridization of DNA duplex, the formation of α-helix and β-sheet peptide structures, and the control of protein folding. This self-assembly strategy has prompted the controlled fabrication of organized structures from synthetic functionalized molecules, colloids, and polymers (so-called supramolecules) [25], where the final structures display additional functionalities in comparison to their "supramolecular building blocks". In addition, hydrogen-bond mediated assembly features several unique attributes [26]. The interaction is highly selective, directional, and with controlled affinity, due to the specifically designed hydrogen donor and acceptor functionalities [27–29]. In addition, the reversible nature of hydrogen-bonding provides not only a self-healing process that helps to reach the thermodynamic equilibrium state but also provides the opportunity to create renewable and stimuli-responsive materials [30].

This article highlights the utilization of hydrogen bonds for the controlled assembly of nanoparticles and polymers both in solution and on substrates. While an extensive discussion of polymer and nanoparticle synthesis is beyond the scope of this article and has been reviewed in other places [31–33], we will discuss here the general aspects pertaining to the design and control of hydrogen-bonding mediated polymer-nanoparticle assemblies and the formation of long-range ordered aggregates. The focus will be directed to the rational chemical design of polymers and nanoparticles, their assembly processes, the control over structural parameters, and the potential practical applications of the resulting composites.

2 Principles of Nanocomposite Design

2.1 Nanoparticles as Building Blocks

Metallic and semiconductor NPs are among the most versatile building blocks employed in the creation of nanoscale materials, which present unique optical, magnetic and electronic properties due to their high surface areas and the confinement of electronic states [9, 34]. Upon assembly, these properties can be readily manipulated according to the size and shape of the nanocomposite materials. Many methods have been reported recently for the synthesis of NPs, including reduction of metal salts in the presence of capping ligands and thermal decomposition of molecular precursors [35–37]. In many cases, the surfaces of NPs are passivated by organic monolayers to protect them from agglomerations; this feature also allows the installation of various chemical functionalities [38–40]. A versatile approach for obtaining monolayer-functionalized NPs involves a two-step process, whereby the NPs are initially synthesized followed by partial or complete exchange of the monolayer with functionalized ligands in a subsequent step. Figure 1 exemplifies this approach for the creation of functionalized gold NPs through solution phase synthesis developed by Brust and coworkers [41, 42], followed by Murray's place-exchange technique [43, 44]. The first step allows control on particle size through the stoichiometry of metal salt to capping ligand. The size of the metallic core can also be modulated based upon the rate of heating for nucleation and the length of the capping ligand with the appropriate solvent. The second step allows incorporation of many types of functionalized thiols into

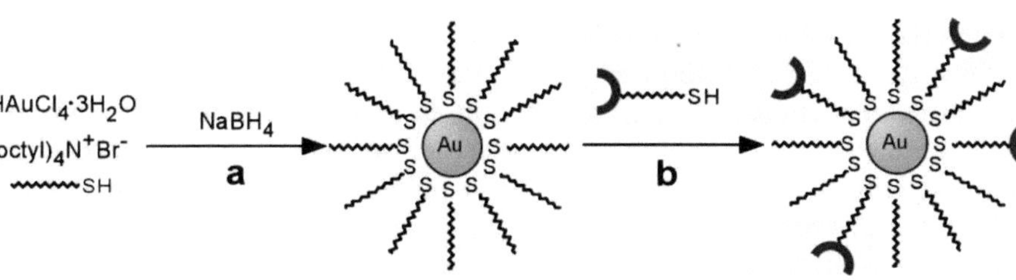

Fig. 1 a The Brust–Schiffrin solution phase gold nanoparticle synthesis, followed by **b** place exchange process developed by Murray

the existing monolayer, as well as the potential for divergent combinatorial methods. Place exchange reactions are not restricted to gold NPs, and have been successfully applied to semiconductor and other metallic (e.g. CdSe [45], Fe_2O_3 [46], FePt [47]) NPs as well.

Characterization of derivatized NPs can be performed using standard solution-phase techniques such as nuclear magnetic resonance (NMR), infrared (IR), and UV-Vis spectroscopy. As will be demonstrated, derivatization of NPs with specifically tailored functional groups allows exploitation of ionic and hydrogen bonding interactions for the creation of controlled assemblies for a variety of applications.

2.2 Polymer Scaffolds

In many cases, the accurate control over length, polydispersity, and functionality of polymers is required for specific applications. Several recent advances in polymer design criteria and synthesis [48–51] allow preparation of well-defined polymers with desired functionalities [52].

To provide the affinity between polymers and other building blocks, the polymer has to be modified with favorably interacting functional groups. There are two methods for the creation of functionalized polymers: polymerization of derivatized monomers, and post-polymerization functionalization. Polymerization of modified monomers requires functionalities that are inert to the polymerization reaction. Anionic polymerization, as the leading technique for controlled synthesis of polymers, is intolerant to the existence of most functionalities. Several recently developed polymerization systems, such as living radical polymerization (LRP) and ring-opening metathesis polymerization (ROMP), offer more versatile alternatives. In LRP, the growing chain is rapidly stabilized by an end group that then dissociates to controlled addition of monomers. Therefore, different monomers can be polymerized in batches to produce structurally distinct block copolymers [48, 49]. In ROMP, a stable metal catalyst breaks a cyclic alkene and the resulting complex keeps reacting with other cyclic alkenes to yield a chain of opened rings. Current catalysts are highly tolerant to various chemical conditions so that a variety of substituted monomers can be prepared and polymerized [51].

Polymerization of modified monomers makes the polymerization itself more challenging, as polymerization parameters known for common monomers, such as copolymerization reaction rates, do not necessarily apply to pre-modified monomers. Post-polymerization functionalization methods, however, enable the use of functionalities as the side-chain modifiers to a well-defined polymer backbone so that a variety of functional polymers can be produced through one single polymer scaffold. A major challenge of this method is that the modification step must be a clean (i.e., free from side products) and quantitative reaction; otherwise, the outcome polymers would be ill defined and afford poor reproducibility of performance between different batches. Shown in Fig. 2 are the reactions that have been reported in recent literature, including nucleophilic substitution [53], Huisgen 1,3-dipolar cycloaddition ("click chemistry") [54–56], and esterification [55].

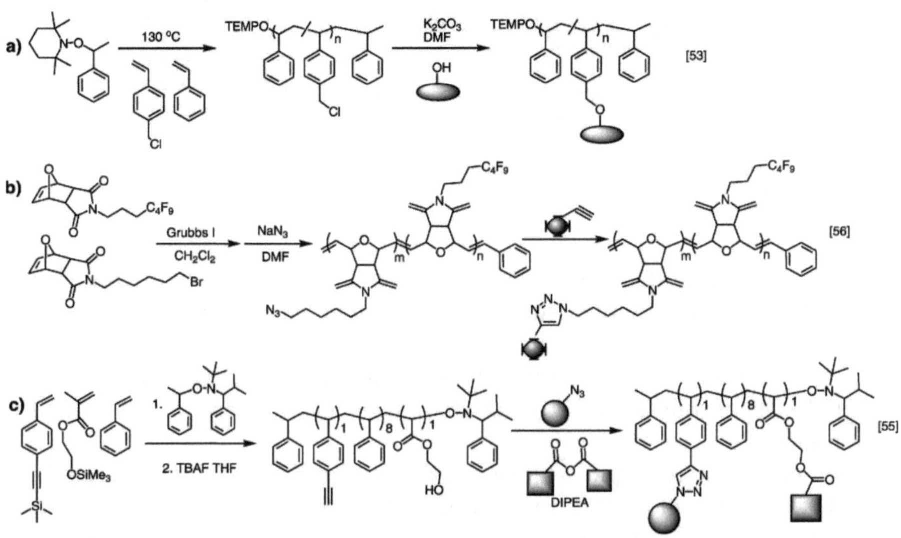

Fig. 2 Examples of polymer synthesis and modification. Several recently developed polymerization systems offer better control of molecular weight and monodispersity: (**a**, **c**) LRP and **b** ROMP. Post-polymerization functionalization reactions include **a** nucleophilic substitution, (**b**, **c**) 1,3-dipolar cycloaddition—"click chemistry," and **c** esterification

Dendrimers are a relatively new class of highly branched, monodispersed polymers [57, 58]. The monodispersity, size-controllable globular structures, and void-containing shapes enable the use of dendrimers as nano building blocks [59, 60] to fabricate advanced functional materials [61] (e.g. catalysis, delivery agents). We will discuss here some examples on the use of dendrimers as supramolecular building blocks to buildup nanocomposite materials (e.g. spacing nanoparticles, or tuning polymer morphologies).

2.3
Surface Modification and Patterning

Utilization of ultra-thin grafted films to impart desired properties onto surfaces has become a powerful technique to prepare composite and functional materials. This modification also includes the tailoring of interfacial interactions by suitable chemical groups to control the adsorption of

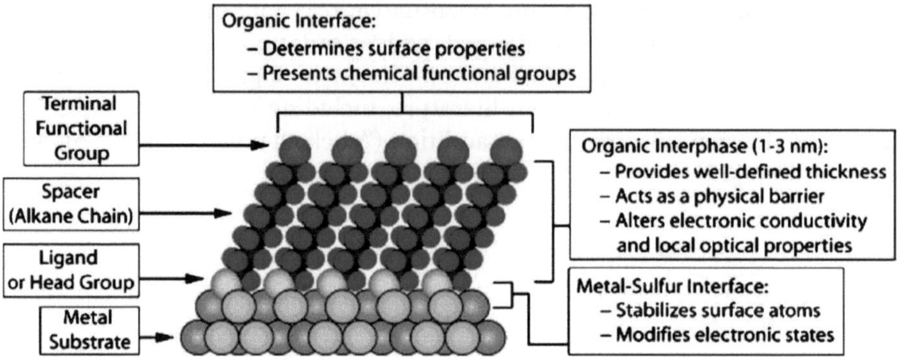

Fig. 3 An ideal SAM of alkanethiols supported on bare gold substrates. Reprinted with permission from [63]

molecules and the stability of resultant films. Self-assembled monolayers (SAMs) [62] provide highly ordered and oriented thin layers that can incorporate a whole variety of functionalities (Fig. 3). Therefore, a wide range of surfaces with specific interactions can be produced with appropriate chemical design. SAMs of functionalized long-chain hydrocarbons are most frequently explored [63].

Engineered microscopic surface structures for site-selective immobilization of NPs or polymers play an essential role in the "bottom-up" fabrication of microelectrodes, biochips, and microfluidic devices. The surface patterning techniques [64] vary from traditional photolithography and scanning electro-beam lithography, to less conventional fabrication approaches, including soft lithography (molding, embossing, and printing) [65], scanning probe lithography (Dip-Pen nanolithography) [66], spreading edge lithography [67, 68], and templated self-assembly (nanosphere [69] and diblock copolymer lithography [70]). Surfaces patterned with modified SAMs and/or polymers can be used to present specific ligands/functionalities with controlled surface densities within well-defined regions. In the following sections, we will discuss some aspects of surface templated two-dimensional organization of nano-building blocks.

3 Control of Structure and Functionality

In this section we will illustrate different ways to control the assembly process, starting from the control over one-dimensional (1D) interparticle distances in bulk aggregates, going to two-dimensional (2D) NP organization using templated polymer films, and ending with the morphologically controlled construction of three-dimensional (3D) aggregates (Fig. 4).

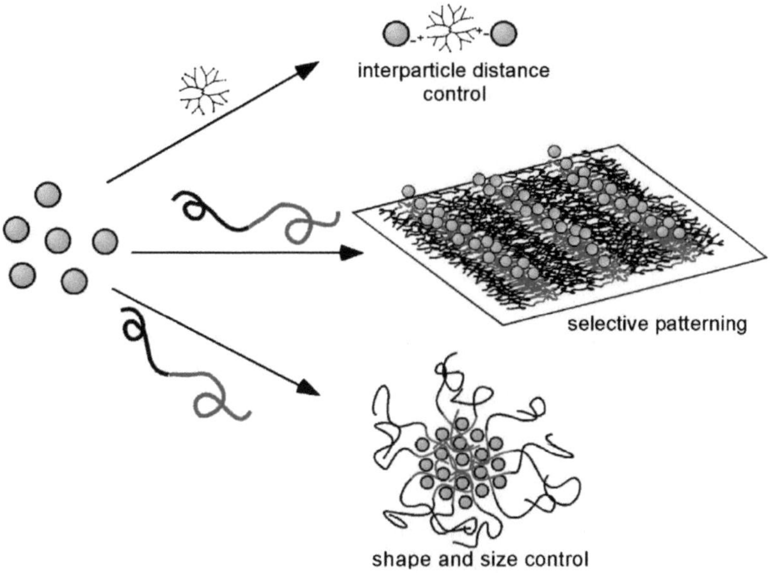

Fig. 4 Schematic illustration of nanoparticle and polymer-based nanocomposites with the control of interparticle spacing for tunable photonic and magnetic properties, site-selective 2D patterning for device fabrication, and size and shape of polymer-particle aggregates for complex 3D structures

3.1
Control of Interparticle Distance

The electronic and magnetic properties of nanoparticles are affected by neighboring particles in a strongly distance-dependent interaction [71, 72]. While control of interparticle distances can be achieved through manipulation of the protecting monolayers of nanoparticles [73–75], the use of separate entities such as polymers/dendrimers to space and regulate the interparticle distance is advantageous as it is more modular and can be done with synthetically accessible building blocks.

Frankamp et al. demonstrated the direct control of interparticle spacing through self-assembly of gold nanoparticles with polyamido(amine) (PAMAM) dendrimers [76]. Gold nanoparticles were functionalized with carboxylic-acid terminal groups to provide the driving force for assembly. Salt bridge formation between the dendrimer surface amine groups and the nanoparticle carboxylic acid groups led to self-assembly between the dendrimer and nanoparticle components. Small angle X-ray scattering (SAXS) revealed a monotonic increase in interparticle spacing from 4.1 to 6.1 nm observed with increasing dendrimer generation (Fig. 5). This assembly resulted in a thin film of gold particles spaced in all directions by PAMAM dendrimers,

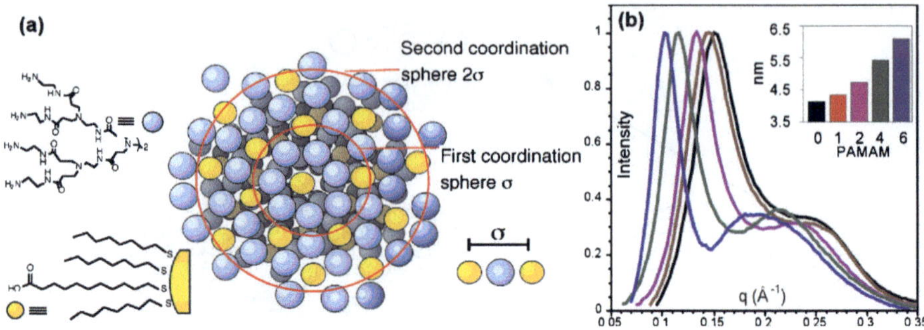

Fig. 5 **a** Nanoparticle assembly using PAMAM dendrimers. **b** Small-angle X-ray scattering plots, demonstrating increasing inter-particle distances with increasing dendrimer generation. Reprinted with permission from [76]

providing a versatile means to systematically control spacing regardless of particle size.

The plasmonic properties of metallic nanoparticles have recently been exploited in optical materials and sensors. In general, the relative wavelength of the surface plasmon resonance (SPR) is dictated by multiple factors including volume, solvent, and dipolar coupling. Dendrimers were used as spacers to control interparticle spacing so as to tailor dipolar coupling and control the optical response [77] through the same self-assembly strategies described above. UV-Vis spectroscopy (Fig. 6) was used to analyze the SPR of each nanaoparticle/PAMAM sample, demonstrating the modulation of dipolar interactions upon assembly. The interparticle spacing was tuned over a 2.1 nm range through choice of dendrimer generation, resulting in an 84 nm SPR red-shift.

Similarly, interparticle spacing is a key determinant of the properties of magnetic particle assemblies [78]. To provide a supramolecular method

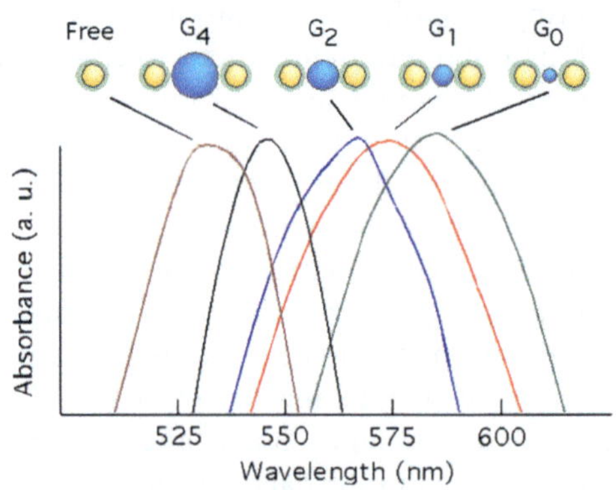

Fig. 6 UV-Vis studies on thin films on nanoparticles assembled with G_0 through G_4 dendrimers and nanoparticles free in solutions. Reprinted with permission from [77]

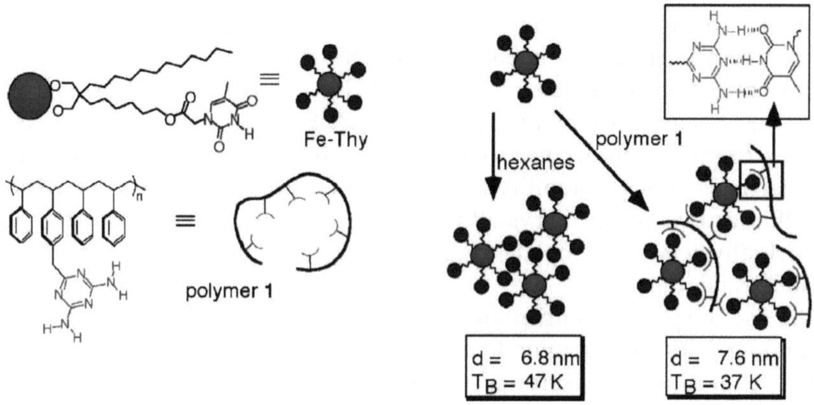

Fig. 7 Illustration depicting molecular recognition based polymer-mediated nanoparticle self-assembly. Interparticle distance and corresponding blocking temperature (T_B) of hexanes induced precipitations and polymer mediated aggregates

for creating controlled interparticle distances, similar "bricks and mortar" methodology [20] was used to assemble γ-Fe_2O_3 nanoparticles. Complementary components for the assembly process were provided by thymine-functionalized iron nanoparticles (Fe-Thy) and triazine-functionalized polymers [79]. Fe-Thy was prepared from $\sim$ 6.5 nm diameter γ-Fe_2O_3 nanoparticles formed using the Alivisatos cupferron method [80]. The monolayer was then modified with a thymine-functionalized diol ligand [46]. γ-Fe_2O_3 nanoparticles are superparamagnets at room temperature, which become ferromagnetic at low temperatures. The superparamagnetic-ferromagnetic transition occurs at a particular temperature, termed as blocking temperature T_B, depending upon the dipolar interactions of particles. As shown in Fig. 7, for the hexane-precipitated particles, a T_B of 47 K was observed; upon assembly with complementary polymers, the T_B decreased to 37 K. This decrease of T_B is a direct consequence of the increased spacing resulting from assembly as opposed to simple precipitation of the nanoparticles. The modulation of magnetic properties was also studied for magnetic nanoparticles assembled with PAMAM dendrimers through electrostatic interactions [81], which is beyond the scope of this review. Controlled spacing of γ-Fe_2O_3 nanoparticles with dendrimers showed a sequential change in blocking temperatures.

3.2 Directed Assembly of Nanobuilding Blocks on Planar Substrates

Self-assembly of nanoparticles in well-ordered 2D arrays represents a major goal in the fabrication of microelectronics devices. Different methods have been developed to tackle the 2D nanoparticle organization challenge.

As shown in Fig. 8, Schimid and coworkers have reported the creation of two different packing arrangements (hexagonal and cubic) of sulfonic acid-functionalized nanoparticles in extended, long-range-ordered arrays using substrates coated with polyethyleneimine (PEI) [82]. The ordering is attributed to the low molecular weight ($M_W = 60$ K) polymer, which allows the nanoparticles enough mobility to rearrange on the surface. It is well known that PEI has two types of crystalline phases, therefore, in this case, the polymer actually acts as a template to direct the arrangement of nanoparticles on the substrate.

The use of specific multi-point hydrogen bonding for selective deposition of nanoparticles on polymer templated surfaces provides a more powerful tool in generating ordered functional arrays [10]. These modifications feature great selectivity due to the involved recognition dyads and provide new directions in terms of renewability based on the reversible nature of hydrogen bonds. Moreover, these surfaces are capable of responding to environmental stimuli, resulting in structural reorganization of the grafted layers and the respective modulation of desired properties.

Binder and coworkers reported the binding of gold nanoparticles onto microphase separated block copolymer films through specific six-point hydrogen bonding interactions [56] (Fig. 9). Gold nanoparticles (5 nm) were coated with ligands consisting of barbituric acid moieties. Block copolymers beared the complementary receptor (Hamilton receptor: Ka = 1.2×10^5 M^{-1}) in one block and a fluorinated side chain in the other block to enhance the microphase separation. Nanoparticles binded only to specific regions, with the size in accordance with the domain size of the block copolymer. In comparison, incubating in solutions containing nanoparticles bearing *N*,*N*-dimethylbarbiturate moieties (relatively small binding constant with Hamil-

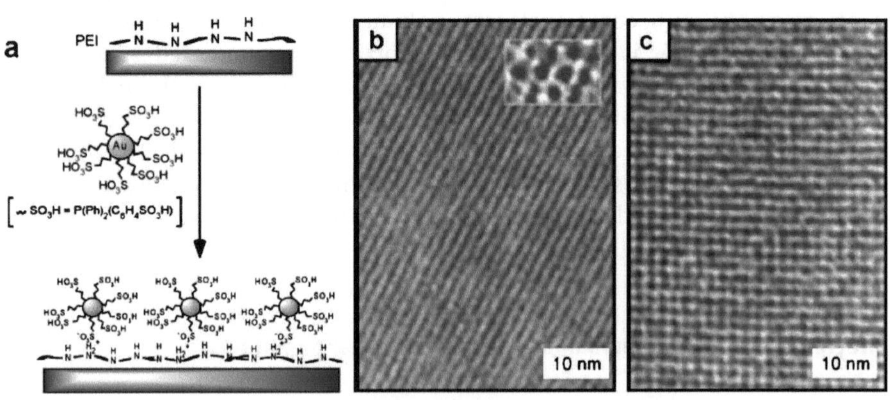

Fig. 8 **a** Schematic illustration of PEI-mediated assembly of gold nanoparticles. **b** Hexagonal and **c** cubic packing arrangements of nanoparticles through transmission electron microscopy (TEM). Reprinted (in part) with permission from [82]

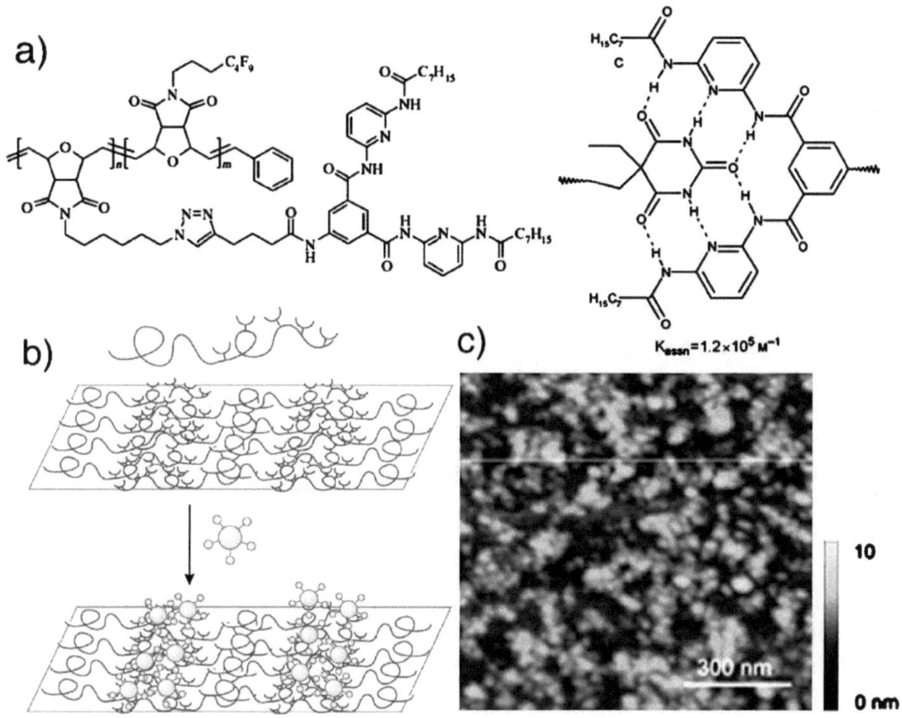

Fig. 9 **a** Chemical structure of the diblock copolymer and the six-point H-bonding. **b** Deposition of NPs onto block copolymer microphase separated film. **c** AFM image indicating the specific deposition of NPs. Reprinted with permission from [56]

ton receptor: Ka = $15\ M^{-1}$) resulted in only nonspecific deposition that did not reflect the microphase separation patterns.

The great selectivity offered by specific supramolecular interactions enable these interactions to be manipulated independently and simultaneously, providing orthogonal self-assembly. This concept, integrated with various microlithographical techniques, provides rapid and site-selective adsorption of molecules and mesoscopic objects at pre-patterned regions, an important issue in many surface-related applications. Rotello and coworkers recently demonstrated orthogonal self-assembly using both hydrogen bonding and electrostatic interactions [83]. Silicon substrates were first patterned with thymine (Thy-PS) and positively charged N-methylpyridinium (PVMP) containing polymers using photolithography. Diaminopyridine-functionalized polystyrene (DAP-PS) and carboxylate derivatized CdSe/ZnS core-shell nanoparticles (COO-NP) were then selectively deposited at the predefined regions of these surfaces. Three-point hydrogen bonding between DAP-PS : Thy-PS and electrostatic interaction between PVMP : COO-NP contribute to the orthogonal self-assembled patterns respectively. This

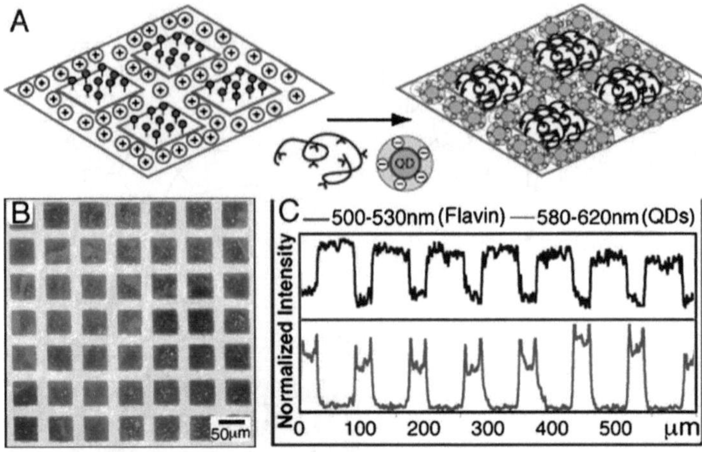

Fig. 10 **A** Orthogonal surface functionalization through Thy-PS:DAP-PS recognition and PVMP:COO-NP electrostatic interactions. **B** Fluorescence microscopy and **C** confocal intensity profile of modified surfaces

recognition-induced simultaneous and independent self-assembly provides high specificity and selectivity in both sequential and one-step functionalization of surfaces (Fig. 10).

3.3 Detailed Tailoring of Self-Organized Structures

Polymeric supramolecular nanostructures allow straightforward tailoring of resulting morphologies and the concurrent switching of functional properties. The specificity and reversibility presented by hydrogen bonding provide the fine-tuning of the resultant morphological structures.

Binke and Ikkala reported the control of microstructures on two length scales, using diblock copolymers consisting of a coil-like polystyrene (PS) block and a poly(4-vinyl pyridine) (P4VP) block, which is stoichiometrically protonated with methane sulfonic acid (MSA) that is then hydrogen bonded to pentadecylphenol (PDP) [84] (Fig. 11a,b). At room temperature, the $P4VP(MSA)_1(PDP)_1$ and PS blocks formed alternating layers with a long period $L_b \approx 35$ nm. In addition, the $P4VP(MSA)_1(PDP)_1$ layers were further microphase separated into another lamellar structure with a period $L_c \approx 4.8$ nm. These mutually perpendicular lamellar structures were confirmed by TEM as shown in Fig. 11d. With increased temperature, hydrogen bonds between PDP and MSA were slowly broken down. Therefore, the lamellar morphology within the $P4VP(MSA)_1(PDP)_1$ layers became disordered first. With the amount of dissociated PDP increasing and becoming miscible with the PS domain, the PS domain swelled and finally the whole morphology was transformed into cylindrical structures (Fig. 11c). The whole transition

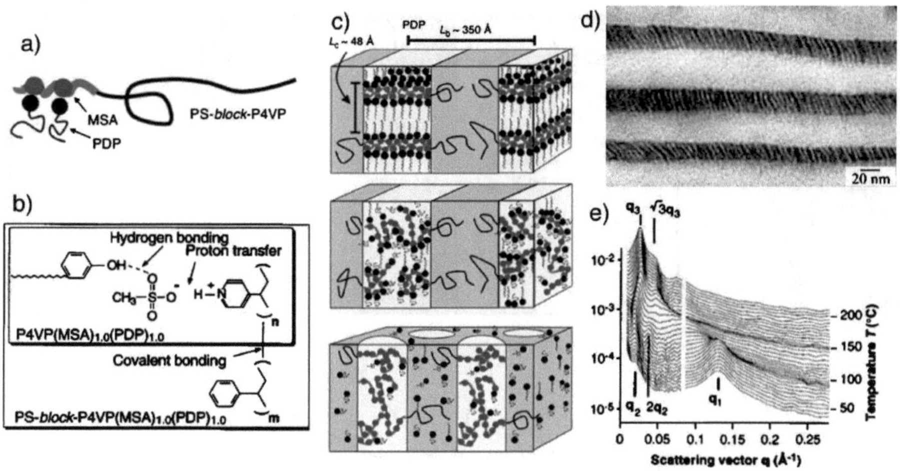

Fig. 11 **a** Schematic illustration and **b** chemical structure of the diblock copolymer. **c** Illustration of the morphology changes with temperature increases; from top to bottom represent three temperature regions: below 100 °C, 100–150 °C, and above 150 °C. **d** TEM image of diblock film at room temperature. (**e**) SAXS intensity curves during heating at 2 °C/min. Reprinted with permission from [84]

process was clearly revealed by SAXS analysis (Fig. 11e). The peaks at q_1 correspond to lamellar order within P4VP(MSA)$_1$(PDP)$_1$ layers. The peaks at q_2 and $2q_2$ represent the long-range lamellar order caused by micro-phase separation of PS-b-P4VP MSA)$_1$(PDP)$_1$, which is replaced by peaks at q_3 and $\surd 3q_3$ near 150 °C, due to the transition to a cylindrical morphology when hydrogen bonding is broken at high temperature.

Shenhar et al. recently demonstrated the selective modification of specific domain properties of block copolymer films. Guest molecules with dendritic substituents of varying size were used to obtain different equilibrium morphologies from a single block copolymer scaffold [85]. Polyether dendrons with different generations nearly double in volume with each increase in generation number (Gx) and perform different abilities in swelling the PS/DAP block, which resulted in a series of morphologies (Lamellar, Cylindrical, and Spherical) from a single PS-b-PS/DAP diblock copolymer (Fig. 12).

3.4 Polymer and Nanoparticle 3D Aggregates in Solution

The noncovalent interactions that define the self-assembly process are responsible for the highly ordered, diverse systems found in nature, providing inspiration for the creation of new self-assembled structures. An inherent property of the bottom-up approach is the ability to achieve a controlled assembly process and to create complex 3D nano-structures. To obtain com-

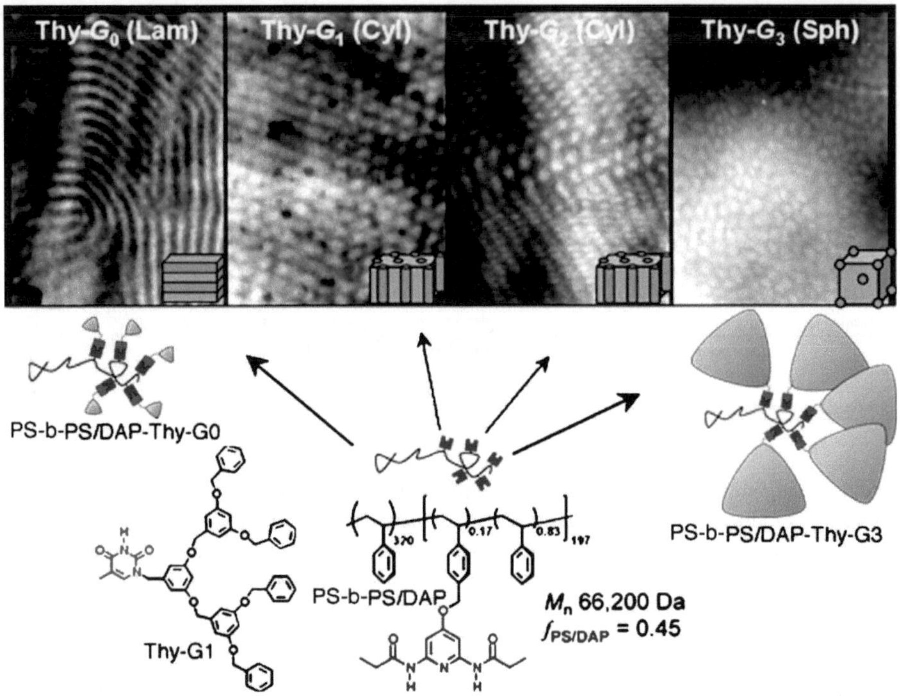

Fig. 12 The specific swelling of PS-b-PS/DAP diblock copolymer using complementary Thy-Gx polyether dendron resulted in a series of morphologies (through TEM analysis)

plete control over the thermodynamics of the assembled system, the careful choice of assembly interactions is the key factor. While electrostatic interactions offer a simple assembly mechanism, they usually provide limited control of the overall aggregate structures due to the far-from-equilibrium conditions, as there is little entropic penalty associated with the large enthalpy gain. The most frequent outcome is extended crosslinked networks arising from kinetic entrapment. However, the thermodynamic control inherent in the recognition process, provided by multi-point H-bonding, expresses itself in the equilibrium-regulated assembly. Assembly of thymine-functionalized gold nanoparticles (Thy-Au) with diaminotriazine (triaz)-functionalized polystyrene in noncompetitive solvents resulted in the creation of huge spherical nanoparticle aggregates, consisting of up to 1.5×10^6 individual nanoparticles (Fig. 13). In contrast, no aggregation was detected when the highly analogous MeThy-functionalized nanoparticles were used instead of Thy-Au, demonstrating the crucial role of specific hydrogen-bonding interactions in the assembly process [20, 86]. The diameter of these aggregates was strongly dependent on the temperature at which the assembly process was performed, ranging from 100 nm to 0.5–1 μm by cooling the system from ambient conditions to – 20 °C.

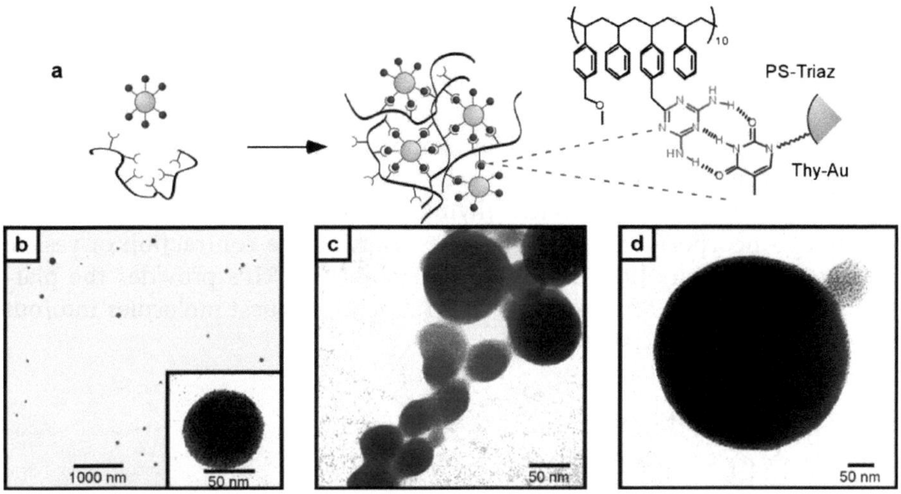

Fig. 13 **a** Schematic illustration of recognition-mediated nanoparticle-polymer assembly. TEM images of Thy-Au/Triaz-PS aggregates formed at **b** 23 °C, **c** 10 °C, and **d** – 20 °C. Reprinted with permission from [20]

The complex thermodynamic assembly process and the resultant temperature-dependent size and shape of the aggregates are actually the main limitation of this method to achieve highly ordered composites. In a follow up investigation, the use of recognition functionalized diblock copolymer (instead of the previous random copolymer) provided not only the formation of tens of nanometer aggregates but also the precise control of the aggregation size by tuning the length of the diblock copolymer [87] (Fig. 14).

Vesicles and liposomes are versatile supramolecular systems with unusual stability and great potential as functional materials with applications in biosensors, targeting agents, microreactors, and encapsulation/drug delivery.

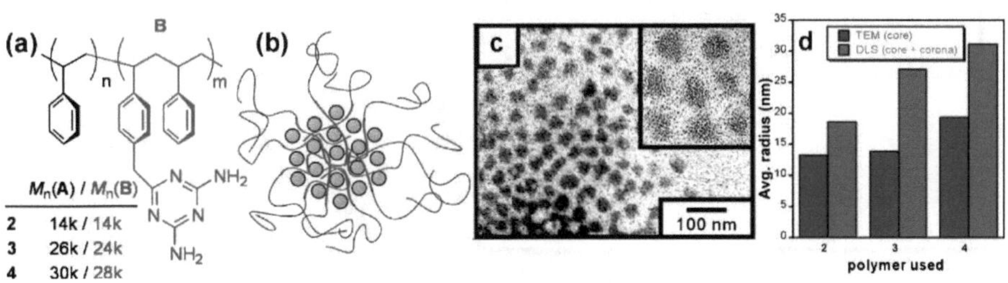

Fig. 14 **a** Chemical structure of PS-b-Tria-PS diblock copolymers used to control the aggregate size. **b** Schematic illustration of aggregated nanoparticles in diblock polymer micelles. **c** Representative TEM images demonstrate the control of aggregate size via the polymer length. **d** Comparison between average core size calculated from TEM images and hydrodynamic radii determined by dynamic light scattering (DLS)

The combination of complementary DAP and Thy functionalized polymers in nonpolar media results spontaneously in the formation of giant vesicular aggregates or recognition-induced polymersomes (RIPs) featuring a 3–5 μm diameter and a ∼ 50 nm membrane thickness [88–90] (Fig. 15). These vesicular structures can be disrupted by complementary small molecules, providing a mechanism for controlled release of materials incorporated within the vesicles. However, multivalent thymine-functionalized nanoparticles can be rapidly incorporated into the vesicles, causing the contraction of vesicle diameters over time [91]. This tailored stability of RIPs provides the platform for the control of attachment and releasing of guest molecules into/out of vesicular polymeric structures.

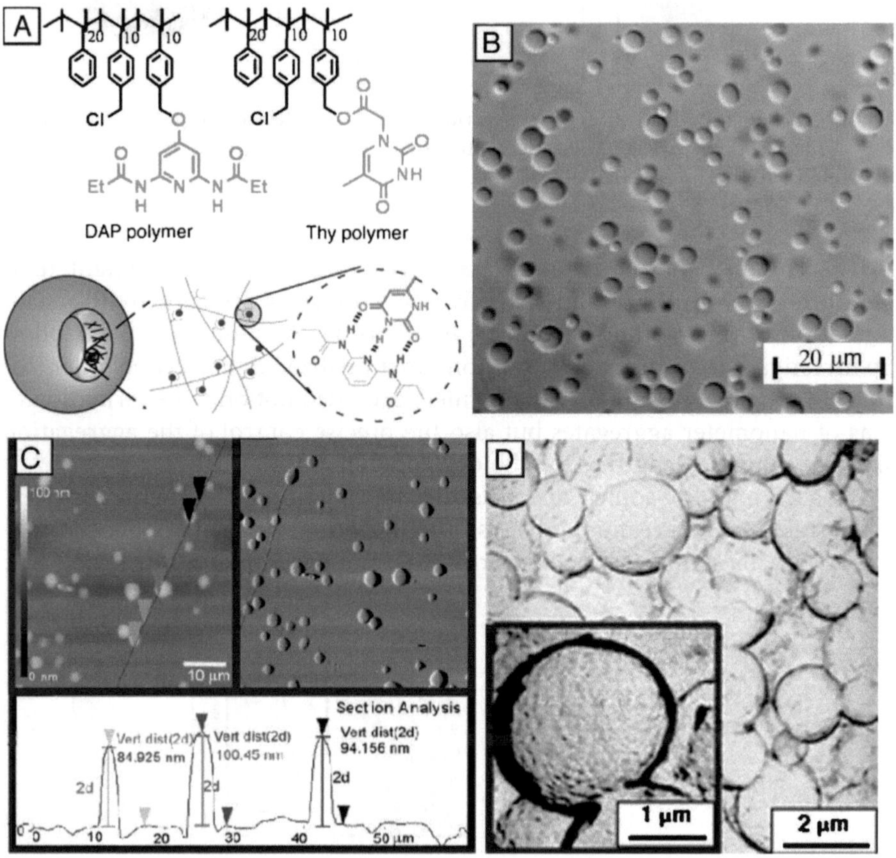

Fig. 15 **A** Chemical structures of Thy- and DAP-functionalized polymers, and schematic illustration of vesicle formation. **B** Differential interference contrast microscopy (DIC), **C** AFM, and **D** fractured TEM images of resultant vesicular aggregates. Reprinted with permission from [88]

4
Conclusion and Outlook

The hydrogen-bond mediated self-assembly of nanoparticles and polymers provides a versatile and effective method to control interparticle distances, assembly shapes, sizes, and anisotropic ordering of the resultant nanocomposites. This approach presents the bottom-up strategy to fabricate nanomaterials from molecular building blocks, which have great potential for assembling and integrating nanoscale materials and particles into advanced structures, systems, and devices.

Great progress has been made in making complex binary lattices and 2D assemblies through bottom-up self-assembly processes. Various nano-scale building blocks are becoming available and new architectures are being developed from them. Nonetheless, the ultimate question might be: how much complexity can be reached through materials self-assembly? Simple nonspecific packing and entropy gain can be insufficient in designing multicomponent complex structures, thereby specific interactions with great directionality and selectivity must be employed. Besides spontaneous self-assembly, directed assembly with an external energy input (e.g. electric or magnetic field) has become a promising option, but was not discussed here. Furthermore, self-assembly guided by lithographically defined surface patterns enables one to integrate the bottom-up assembly strategy with well-established top-down lithographic techniques. This integration has begun to attract more and more attention for producing nanostructures with long-range ordering and large-scale production.

The true challenge for self-assembly goes far beyond the prediction and manipulation of complex structures: the real aim is the design and synthesis of building blocks that, once assembled into ordered structures, will produce certain specific functions—"form is function". Because of the diverse nature of polymers and nanoparticles that allow the incorporation of specific functionalities (e.g. mechanical properties, optical emission, electrical conductivity, biocompatibility, and magnetism), it is convenient to construct functions into newly formed materials.

The examples given in this review represent proof-of-concept demonstrations of using hydrogen-bonding mediated assembly to create ordered nanocomposite materials, although there are still some technical barriers and theoretical challenges to be overcome. Key requirements for future developments in this area might include: (a) advances in the controlled synthesis of building blocks; (b) deep understanding of the intricate balance between entropy and enthalpy gain/loss during assembly; and (c) integration of various assembly routes in an orthogonal or combinatory fashion to obtain desired hierarchical structures. The complexity and the resulting utility of polymer-nanoparticle-based nanocomposites will keep increasing along the way as we gain better understanding and control of the overall self-assembly process.

References

1. Bhat RR, Genzer J, Chaney BN, Sugg HW, Liebmann-Vinson A (2003) Nanotechnology 14:1145
2. Shenhar R, Norsten TB, Rotello VM (2005) Adv Mater 17:657
3. Haryono A, Binder WH (2006) Small 2:600
4. Binder WH (2005) Angew Chem Int Ed 44:5172
5. Giannelis EP (1996) Adv Mater 8:29
6. Sanchez C, Soler-Illia G, Ribot F, Lalot T, Mayer CR, Cabuil V (2001) Chem Mater 13:3061
7. Fendler JH (1996) Chem Mat 8:1616
8. Ozin GA, Arsenault AC (2005) Nanochemistry: A Chemical Approach to Nanomaterials. Royal Society of Chemistry, Cambridge
9. Collier CP, Vossmeyer T, Heath JR (1998) Annu Rev Phys Chem 49:371
10. Murray CB, Kagan CR, Bawendi MG (2000) Annu Rev Mater Sci 30:545
11. Alivisatos AP, Johnsson KP, Peng XG, Wilson TE, Loweth CJ, Bruchez MP, Schultz PG (1996) Nature 382:609
12. Winfree E, Liu FR, Wenzler LA, Seeman NC (1998) Nature 394:539
13. Gittins DI, Bethell D, Schiffrin DJ, Nichols RJ (2000) Nature 408:67
14. Whaley SR, English DS, Hu EL, Barbara PF, Belcher AM (2000) Nature 405:665
15. Cui Y, Lieber CM (2001) Science 291:851
16. Huang Y, Duan XF, Cui Y, Lauhon LJ, Kim KH, Lieber CM (2001) Science 294:1313
17. Shimomura M, Sawadaishi T (2001) Curr Opin Colloid Interface Sci 6:11
18. Liu ST, Maoz R, Schmid G, Sagiv J (2002) Nano Lett 2:1055
19. Zhang SG (2003) Nat Biotech 21:1171
20. Boal AK, Ilhan F, DeRouchey JE, Thurn-Albrecht T, Russell TP, Rotello VM (2000) Nature 404:746
21. Shevchenko EV, Talapin DV, Kotov NA, O'Brien S, Murray CB (2006) Nature 439:55
22. Boncheva M, Whitesides GM (2005) MRS Bull 30:736
23. Hamley IW (2003) Angew Chem Int Ed 42:1692
24. Lehn JM (1993) Science 260:1762
25. Binder WH (2005) Monatsh Chem 136:1
26. Lehn JM (2002) Science 295:2400
27. Prins LJ, Reinhoudt DN, Timmerman P (2001) Angew Chem Int Ed 40:2382
28. Beijer FH, Kooijman H, Spek AL, Sijbesma RP, Meijer EW (1998) Angew Chem Int Ed Engl 37:75
29. Sivakova S, Rowan SJ (2005) Chem Soc Rev 34:9
30. Xu H, Norsten TB, Uzun O, Jeoung E, Rotello VM (2005) Chem Commun: 5157
31. Trindade T, O'Brien P, Pickett NL (2001) Chem Mater 13:3843
32. Daniel M-C, Astruc D (2004) Chem Rev 104:293
33. Percec V (2001) Chem Rev 101:3579
34. El-Sayed MA (2001) Acc Chem Res 34:257
35. Goia DV (2004) J Mater Chem 14:451
36. Esteves ACC, Trindade T (2002) Curr Opin Solid State Mater Sci 6:347
37. John VT, Simmons B, McPherson GL, Bose A (2002) Curr Opin Colloid Interface Sci 7:288
38. Hostetler MJ, Templeton AC, Murray RW (1999) Langmuir 15:3782
39. Drechsler UBE, Rotello VM (2004) Chem Eur J 10:5570
40. Shenhar R, Rotello VM (2003) Acc Chem Res 36:549

41. Brust M, Walker M, Bethell D, Schiffrin DJ, Whyman R (1994) J Chem Soc, Chem Commun: 801
42. Brust M, Fink J, Bethell D, Schiffrin DJ, Kiely C (1995) J Chem Soc Chem Commun: 1655
43. Templeton AC, Wuelfing WP, Murray RW (2000) Acc Chem Res 33:27
44. Templeton AC, Hostetler MJ, Warmoth EK, Chen S, Hartshorn CM, Krishnamurthy VM, Forbes MDE, Murray RW (1998) J Am Chem Soc 120:4845
45. Peng XG, Wilson TE, Alivisatos AP, Schultz PG (1997) Angew Chem Int Ed Engl 36:145
46. Boal AK, Das K, Gray M, Rotello VM (2002) Chem Mater 14:2628
47. Hong R, Fischer NO, Emrick T, Rotello VM (2005) Chem Mater 17:4617
48. Hawker CJ, Bosman AW, Harth E (2001) Chem Rev 101:3661
49. Coessens V, Pintauer T, Matyjaszewski K (2001) Prog Polym Sci 26:337
50. Chiefari J, Chong YK, Ercole F, Krstina J, Jeffery J, Le TPT, Mayadunne RTA, Meijs GF, Moad CL, Moad G, Rizzardo E, Thang SH (1998) Macromolecules 31:5559
51. Bielawski CW, Grubbs RH (2000) Angew Chem Int Ed 39:2903
52. Hawker CJ, Wooley KL (2005) Science 309:1200
53. Shenhar R, Sanyal A, Uzun O, Nakade H, Rotello VM (2004) Macromolecules 37:4931
54. Carroll JB, Jordan BJ, Xu H, Erdogan B, Lee L, Cheng L, Tiernan C, Cooke G, Rotello VM (2005) Org Lett 7:2551
55. Malkoch M, Thibault RJ, Drockenmuller E, Messerschmidt M, Voit B, Russell TP, Hawker CJ (2005) J Am Chem Soc 127:14942
56. Binder WH, Kluger C, Straif CJ, Friedbacher G (2005) Macromolecules 38:9405
57. Matthews OA, Shipway AN, Stoddart JF (1998) Prog Polym Sci 23:1
58. Jikei M, Kakimoto M (2001) Prog Polym Sci 26:1233
59. Zimmerman SC, Zeng FW, Reichert DEC, Kolotuchin SV (1996) Science 271:1095
60. Schenning A, Elissen-Roman C, Weener JW, Baars M, van der Gaast SJ, Meijer EW (1998) J Am Chem Soc 120:8199
61. Vogtle F, Gestermann S, Hesse R, Schwierz H, Windisch B (2000) Prog Polym Sci 25:987
62. Ulman A (1996) Chem Rev 96:1533
63. Love JC, Estroff LA, Kriebel JK, Nuzzo RG, Whitesides GM (2005) Chem Rev 105: 1103
64. Gates BD, Xu Q, Stewart M, Ryan D, Willson CG, Whitesides GM (2005) Chem Rev 105:1171
65. Xia YN, Whitesides GM (1998) Angew Chem Int Ed 37:551
66. Piner RD, Zhu J, Xu F, Hong SH, Mirkin CA (1999) Science 283:661
67. Aizenberg J, Black AJ, Whitesides GM (1999) Nature 398:495
68. Aizenberg J, Black AJ, Whitesides GM (1998) Nature 394:868
69. Haynes CL, Van Duyne RP (2001) J Phys Chem B 105:5599
70. Park M, Harrison C, Chaikin PM, Register RA, Adamson DH (1997) Science 276:1401
71. Taton TA, Mirkin CA, Letsinger RL (2000) Science 289:1757
72. Sandrock ML, Foss CA (1999) J Phy Chem B 103:11398
73. Norsten TB, Frankamp BL, Rotello VM (2002) Nano Lett 2:1345
74. Park S-J, Lazarides AA, Storhoff JJ, Pesce L, Mirkin CA (2004) J Phys Chem B 108: 12375
75. Ohno K, Koh K-m, Tsujii Y, Fukuda T (2002) Macromolecules 35:8989
76. Frankamp BL, Boal AK, Rotello VM (2002) J Am Chem Soc 124:15146
77. Srivastava S, Frankamp BL, Rotello VM (2005) Chem Mater 17:487
78. Tartaj P, Serna CJ (2002) Chem Mater 14:4396

79. Boal AK, Frankamp BL, Uzun O, Tuominen MT, Rotello VM (2004) Chem Mater 16:3252
80. Rockenberger J, Scher EC, Alivisatos AP (1999) J Am Chem Soc 121:11595
81. Frankamp BL, Boal AK, Tuominen MT, Rotello VM (2005) J Am Chem Soc 127:9731
82. Schmid G, Baumle M, Beyer N (2000) Angew Chem Int Ed 39:181
83. Xu H, Hong R, Lu T, Uzun O, Rotello VM (2006) J Am Chem Soc 128:3162
84. Ruokolainen J, Makinen R, Torkkeli M, Makela T, Serimaa R, Brinke Gt, Ikkala O (1998) Science 280:557
85. Shenhar R, Xu H, Frankamp BL, Mates TE, Sanyal A, Uzun O, Rotello VM (2005) J Am Chem Soc 127:16318
86. Boal AK, Gray M, Ilhan F, Clavier GM, Kapitzky L, Rotello VM (2002) Tetrahedron 58:765
87. Frankamp BL, Uzun O, Ilhan F, Boal AK, Rotello VM (2002) J Am Chem Soc 124:892
88. Uzun O, Xu H, Jeoung E, Thibault RJ, Rotello VM (2005) Chem Eur J 11:6916
89. Ilhan F, Galow TH, Gray M, Clavier G, Rotello VM (2000) J Am Chem Soc 122:5895
90. Uzun O, Sanyal A, Nakade H, Thibault RJ, Rotello VM (2004) J Am Chem Soc 126:14773
91. Thibault RJ, Galow TH, Turnberg EJ, Gray M, Hotchkiss PJ, Rotello VM (2002) J Am Chem Soc 124:15249

Author Index Volumes 201–207

Author Index Volumes 1–100 see Volume 100
Author Index Volumes 101–200 see Volume 200

Anwander, R. see Fischbach, A.: Vol. 204, pp. 155–290.
Ayres, L. see Löwik D. W. P. M.: Vol. 202, pp. 19–52.

Binder, W. H. and *Zirbs, R.*: Supramolecular Polymers and Networks with Hydrogen Bonds in the Main- and Side-Chain. Vol. 207, pp. 1–78
Bouteiller, L.: Assembly via Hydrogen Bonds of Low Molar Mass Compounds into Supramolecular Polymers. Vol. 207, pp. 79–112
Boutevin, B., *David, G.* and *Boyer, C.*: Telechelic Oligomers and Macromonomers by Radical Techniques. Vol. 206, pp. 31–135
Boyer, C., see Boutevin B: Vol. 206, pp. 31–135
ten Brinke, G., *Ruokolainen, J.* and *Ikkala, O.*: Supramolecular Materials Based On Hydrogen-Bonded Polymers. Vol. 207, pp. 113–177

Csetneki, I., see Filipcsei G: Vol. 206, pp. 137–189

David, G., see Boutevin B: Vol. 206, pp. 31–135
Deming T. J.: Polypeptide and Polypeptide Hybrid Copolymer Synthesis via NCA Polymerization. Vol. 202, pp. 1–18.
Donnio, B. and *Guillon, D.*: Liquid Crystalline Dendrimers and Polypedes. Vol. 201, pp. 45–156.

Elisseeff, J. H. see Varghese, S.: Vol. 203, pp. 95–144.

Ferguson, J. S., see Gong B: Vol. 206, pp. 1–29
Filipcsei, G., *Csetneki, I.*, *Szilágyi, A.* and *Zrínyi, M.*: Magnetic Field-Responsive Smart Polymer Composites. Vol. 206, pp. 137–189
Fischbach, A. and *Anwander, R.*: Rare-Earth Metals and Aluminum Getting Close in Ziegler-type Organometallics. Vol. 204, pp. 155–290.
Fischbach, C. and *Mooney, D. J.*: Polymeric Systems for Bioinspired Delivery of Angiogenic Molecules. Vol. 203, pp. 191–222.
Freier T.: Biopolyesters in Tissue Engineering Applications. Vol. 203, pp. 1–62.
Friebe, L., *Nuyken, O.* and *Obrecht, W.*: Neodymium Based Ziegler/Natta Catalysts and their Application in Diene Polymerization. Vol. 204, pp. 1–154.

García A. J.: Interfaces to Control Cell-Biomaterial Adhesive Interactions. Vol. 203, pp. 171–190.

Gong, B., *Sanford, AR.* and *Ferguson, JS.*: Enforced Folding of Unnatural Oligomers: Creating Hollow Helices with Nanosized Pores. Vol. 206, pp. 1–29
Guillon, D. see Donnio, B.: Vol. 201, pp. 45–156.

Harada, A., *Hashidzume, A.* and *Takashima, Y.*: Cyclodextrin-Based Supramolecular Polymers. Vol. 201, pp. 1–44.
Hashidzume, A. see Harada, A.: Vol. 201, pp. 1–44.
Heinze, T., *Liebert, T.*, *Heublein, B.* and *Hornig, S.*: Functional Polymers Based on Dextran. Vol. 205, pp. 199–291.
Heßler, N. see Klemm, D.: Vol. 205, pp. 57–104.
Van Hest J. C. M. see Löwik D. W. P. M.: Vol. 202, pp. 19–52.
Heublein, B. see Heinze, T.: Vol. 205, pp. 199–291.
Hornig, S. see Heinze, T.: Vol. 205, pp. 199–291.
Hornung, M. see Klemm, D.: Vol. 205, pp. 57–104.

Ikkala, O., see ten Brinke, G.: Vol. 207, pp. 113–177

Jaeger, W. see Kudaibergenov, S.: Vol. 201, pp. 157–224.
Janowski, B. see Pielichowski, K.: Vol. 201, pp. 225–296.

Kataoka, K. see Osada, K.: Vol. 202, pp. 113–154.
Klemm, D., *Schumann, D.*, *Kramer, F.*, *Heßler, N.*, *Hornung, M.*, *Schmauder H.-P.* and *Marsch, S.*: Nanocelluloses as Innovative Polymers in Research and Application. Vol. 205, pp. 57–104.
Klok H.-A. and *Lecommandoux, S.*: Solid-State Structure, Organization and Properties of Peptide—Synthetic Hybrid Block Copolymers. Vol. 202, pp. 75–112.
Kosma, P. see Potthast, A.: Vol. 205, pp. 151–198.
Kosma, P. see Rosenau, T.: Vol. 205, pp. 105–149.
Kramer, F. see Klemm, D.: Vol. 205, pp. 57–104.
Kudaibergenov, S., *Jaeger, W.* and *Laschewsky, A.*: Polymeric Betaines: Synthesis, Characterization, and Application. Vol. 201, pp. 157–224.

Laschewsky, A. see Kudaibergenov, S.: Vol. 201, pp. 157–224.
Lecommandoux, S. see Klok H.-A.: Vol. 202, pp. 75–112.
Li, S., see Li W: Vol. 206, pp. 191–210
Li, W. and *Li, S.*: Molecular Imprinting: A Versatile Tool for Separation, Sensors and Catalysis. Vol. 206, pp. 191–210
Liebert, T. see Heinze, T.: Vol. 205, pp. 199–291.
Löwik, D. W. P. M., *Ayres, L.*, *Smeenk, J. M.*, *Van Hest J. C. M.*: Synthesis of Bio-Inspired Hybrid Polymers Using Peptide Synthesis and Protein Engineering. Vol. 202, pp. 19–52.

Marsch, S. see Klemm, D.: Vol. 205, pp. 57–104.
Mooney, D. J. see Fischbach, C.: Vol. 203, pp. 191–222.

Nishio Y.: Material Functionalization of Cellulose and Related Polysaccharides via Diverse Microcompositions. Vol. 205, pp. 1–55.
Njuguna, J. see Pielichowski, K.: Vol. 201, pp. 225–296.
Nuyken, O. see Friebe, L.: Vol. 204, pp. 1–154.

Obrecht, W. see Friebe, L.: Vol. 204, pp. 1–154.
Osada, K. and *Kataoka, K.*: Drug and Gene Delivery Based on Supramolecular Assembly of PEG-Polypeptide Hybrid Block Copolymers. Vol. 202, pp. 113–154.

Pielichowski, J. see Pielichowski, K.: Vol. 201, pp. 225–296.
Pielichowski, K., Njuguna, J., Janowski, B. and *Pielichowski, J.*: Polyhedral Oligomeric Silsesquioxanes (POSS)-Containing Nanohybrid Polymers. Vol. 201, pp. 225–296.
Pompe, T. see Werner, C.: Vol. 203, pp. 63–94.
Potthast, A., Rosenau, T. and *Kosma, P.*: Analysis of Oxidized Functionalities in Cellulose. Vol. 205, pp. 151–198.
Potthast, A. see Rosenau, T.: Vol. 205, pp. 105–149.

Rosenau, T., Potthast, A. and *Kosma, P.*: Trapping of Reactive Intermediates to Study Reaction Mechanisms in Cellulose Chemistry. Vol. 205, pp. 105–149.
Rosenau, T. see Potthast, A.: Vol. 205, pp. 151–198.
Rotello, V. M., see Xu, H.: Vol. 207, pp. 179–198
Ruokolainen, J., see ten Brinke, G.: Vol. 207, pp. 113–177

Salchert, K. see Werner, C.: Vol. 203, pp. 63–94.
Sanford, A. R., see Gong B: Vol. 206, pp. 1–29
Schlaad H.: Solution Properties of Polypeptide-based Copolymers. Vol. 202, pp. 53–74.
Schmauder H.-P. see Klemm, D.: Vol. 205, pp. 57–104.
Schumann, D. see Klemm, D.: Vol. 205, pp. 57–104.
Smeenk, J. M. see Löwik D. W. P. M.: Vol. 202, pp. 19–52.
Srivastava, S., see Xu, H.: Vol. 207, pp. 179–198
Szilágyi, A., see Filipcsei G: Vol. 206, pp. 137–189

Takashima, Y. see Harada, A.: Vol. 201, pp. 1–44.

Varghese, S. and *Elisseeff, J. H.*: Hydrogels for Musculoskeletal Tissue Engineering. Vol. 203, pp. 95–144.

Werner, C., Pompe, T. and *Salchert, K.*: Modulating Extracellular Matrix at Interfaces of Polymeric Materials. Vol. 203, pp. 63–94.

Xu, H., Srivastava, S., and *Rotello, V. M.*: Nanocomposites Based on Hydrogen Bonds. Vol. 207, pp. 179–198

Zhang, S. see Zhao, X.: Vol. 203, pp. 145–170.
Zhao, X. and *Zhang, S.*: Self-Assembling Nanopeptides Become a New Type of Biomaterial. Vol. 203, pp. 145–170.
Zirbs, R., see Binder, W. H.: Vol. 207, pp. 1–78
Zrínyi, M., see Filipcsei G: Vol. 206, pp. 137–189

Subject Index

AC orientational switching 162
Aggregates, 3D in solution 191
Alivisatos cupferron method 187
Alkanethiols, SAM 184
Alkylphenols 52
4-Alkyl-resorcinoles 59
Aminopyrazolones 46
Archimedian tiling 126

Barbituric acid 188
Benzene-tricarboxamide (BTC) 83, 102
Bisphenol-A/tetrapyridine 91
Bis-urea 85
Block copolymers, blends, hydrogen bonding 118, 125, 137
Branches 98
Brust–Schiffrin solution phase gold nanoparticle synthesis 181
BTC derivatives 102

Calix[4]arenes 46
Camphorsulfonic acid (CSA) 154
Carboxylate derivatized CdSe/ZnS core-shell nanoparticles (COO-NP) 189
Chain polarity 100
Chain stoppers 101
Chirality 99
p-Chloromethylstyrene 62
Click chemistry 183
Cole–Cole plot 89
Comb copolymers, block copolymer-based hydrogen-bonded 137
–, hydrogen-bonded 129, 135
–, –, mesogenic side chains 157
Conductivity, host polymer's crystallization (polyethylene oxide) 141
–, protonic, tridirectional 143
Copolymers 97
Covalent capture 102
Crosslinks 98
Cyclohexane-tricarboxamide (CTC) 84

DBSA/P4VP-PS 56
DDA–AAD 49
Dialkyl ureas 81, 100
Diaminopyridine functionalized polystyrene (DAP-PS) 189
Diblock copolymer lithography 184
Dilute solutions 128
Dip-Pen nanolithography 184
DNA ligase 102
Dodecylbenzenesulfonic acid (DBSA) 55, 124
Dry brush regime 118

Elastic materials 92
Electroactive polymers 141
Electro-optical properties 100
2-Ethylhexyl-methacrylates 62

Fe_2O_3 nanoparticles 187
Fiber formation 91
Fluorescence spectroscopy 105
FTIR spectroscopy 105
Fullerenated poly(2-hydroxyethyl methacrylate) 11

Glasses, amorphous 91

Hamilton receptor/barbituric acid 123
HBSP solutions, rheological properties 82
HBSPs, bulk 91
HBSPs, macroscopic properties 82
4-Hexylresorcinol (Hres) 154
Hollow particles, PS-*b*-P2VP 136
–, self-assembly 134
Huisgen 1,3-dipolar cycloaddition 183

Hydrogen bonding assembly, layer-by-layer 164
Hydrogen bonds 1, 5
–, monovalent 10
–, surfaces 63
Hydrogen-bonded block copolymers 118
Hydrogen-bonded comb copolymers 129
–, conjugated polymer-based 152
–, homopolymer-based 130
–, mesogenic side chains 157
Hydrogen-bonded interpenetrating polymer networks 167
Hydrogen-bonded polymers, main chain 9
Hydrogen-bonded side-chain LC copolymers, homopolymer-based 157
Hydrogen-bonded supramolecules, polymer-based 118

Interparticle distance, control 185
Interpenetrating polymer networks, hydrogen-bonded 167
Ionic conductivity 141
IPNs, hydrogen-bonded 167
Isothermal titration calorimetry (ITC) 106

Lamellar-*in*-lamellar self-assembly 138
Layer-by-layer hydrogen bonding assembly 164
LC copolymers, side-chain, hydrogen-bonded, block copolymer-based 161
–, homopolymer-based 157
Light scattering 104
Liposomes 193
Liquid crystallinity 10, 93

Mark–Houwink calibration curve 104
Membranes, nanoporous 146
Mesogens 157
Methane sulfonic acid (MSA) 133, 190
Molar mass 102
Monolayers, self-assembled 184
Multilayer thin films, layer-by-layer deposition 118

Nanobuilding blocks, planar substrates 187
Nanocomposites 179
–, design 181
Nano-objects 150
Nanoparticles, building blocks 179, 181
Nanoporous membranes 146
Nanosphere lithography 184
Naphthyridines 50
NMR spectroscopy 105
Nonadecylphenol (NDP) 52, 130, 137
Nonmesogenic side chains 129

Octyl gallate (OG) 155
Oligopeptides 90
Order–disorder transition 130, 138
4-Oxybenzoic acid 157

P4VP 120
–, 5,7-dodecadiynedioic acid hydrogen bonded 137
P4VP-PS 55
PB-*b*-PEO 59
PEK 122
Pentadecylphenol (PDP) 52, 130, 137, 190
Pentamethylene spacer group 157
PEO 132, 165
–/DBSA 59
Photonic bandgap materials 144
–, temperature-dependent 161
PMMA/PEO 11
Poly(acetoxystyrene)/poly(ethyleneoxide) 11
Poly(acetylenes) 13
Poly(acrylamide) 165
Poly(acrylic acid) (PAA) 157
–/poly(*N*,*N*-dimethylacrylamide) 11
–/poly(ethyleneoxide) 11
Poly(alkyl thiophene)s 154
Poly(amic acid) (PAE), carboxyl-terminated 135
Poly(aniline) 58
Poly(butadienes) 13, 125
Poly(*tert*-butyl methacrylate) 125
Poly(2,6-dimethyl-1,4-diphenyl oxide) (PPE) 151
Poly(dimethylsiloxane) 13
Poly(ether ketone) (PEK) 122
Poly(ethylene oxide) (PEO) 132, 165
Poly(ethylene oxide)-*block*-poly(ethylene imine) (PEO-*b*-PEI) 140
Poly(ethylene/butylene) 93, 124
Poly(isobutylenes) (PIB) 13, 122

Poly(isoprene-*block*-2-vinylpyridine) 126
Poly(methyl methacrylate) (PMMA) 124
Poly(methylsilsesquioxane) (PSQ) 167
Poly(*N*-isopropylacrylamide) (PNIPAM) 168
Poly(oxynorbornenes) 60
Poly(*p*-phenylene)s 154
Poly(phenylquinoline)-*block*-polystyrene 135
Poly(2,5-pyridine diyl) 155
Poly(2,5-pyridinium methane sulfonates) 54
Poly(siloxanes) 13
Poly(styrene-*block*-4-hydroxystyrene) 126
Poly(styrene-co-4-vinylphenol) 11
Poly(styrene-co-4-vinylpyridine) 11
Poly(4-trimethylsilylstyrene) 122
Poly(vinyl alcohol) (PVA) 165
Poly(1-vinylimidazole) 11
Poly(1-vinylimidazole)–alkanoic acid 132
Poly(vinylphosphonic acid) 12
Poly(4-vinylpyridine) (P4VP) 11, 120, 190
–, (amphiphilic) phenolic moieties 52
Poly(2-vinylpyridine-*b*-isoprene) 54
Poly(2-vinylpyridine-*b*-isoprene-*b*-2-vinylpyridine) 126
Poly(vinylpyrrolidone) (PVPON) 165
Polyamido(amine) (PAMAM) dendrimers 185
Polyaniline (PANI) 154
Polybutadiene 4
Polyelectrolyte–surfactant complexes 129
Polyether dendrons 191
Polyethyleneimine (PEI) 188
Polyimide (PI), carboxyl-terminated 135
Polymer blends, via H-bonding 10
Polymer conductors, protonic 141
Polymer scaffolds 182
Polymeric material 1
Polymers 179
–, bivalent hydrogen bonds 12
–, multiple hydrogen bonds 39
–, quadruple hydrogen bonds 28
–, side-chain hydrogen-bonded 51
–, trivalent hydrogen bonds 18
Polymersomes, recognition-induced 194
Polystyrene (PS) 122, 190
–, 2,6-diaminopyridine (DAP-PS) 62, 189
–, diaminotriazine (triaz)-functionalized 192
–, monocarboxy-terminated (CPS) 134
Polystyrene-*b*-poly(1,2-butadiene)-*b*-poly(*tert*-butyl methacrylate) 125
Polystyrene-*b*-poly(2-cinnamoyloxyethyl methacrylate) (PS-*b*-PCEMA) 128
Proton conductors 141
Protonic conductivity, tridirectional 143
PS 62, 122, 190
PS(OH)/PCL 11
–/PMMA 11
PS-PVP(PDP)x 55
PS-*b*-P2VP 126
PS-*b*-P4VP diblock 56
PS-*b*-PAA(LC) 164
PS-*b*-PMAA, hydrogen bonding mesogen 161
PS-*b*-PS/DAP diblock 192
P*t*BA-*b*-PCEMA 128
PVMP 189
PVP-PS 52

Recognition-induced polymersomes 194
Reversible volume transitions 167
Ring-chain equilibrium 97

Scanning probe lithography (Dip-Pen nanolithography) 184
Self-assembled monolayers 184
Self-assembly 1
–, bulk state 130, 137
–, dilute solution, hollow spheres 134
–, lamellar-*in*-lamellar 138
–, monolayers 184
–, templated 184
Self-organized structures, tailoring 190
Side-chain hydrogen-bonded polymers 51
Side-chain liquid-crystalline (LC) block copolymers 146
- polymers, hydrogen-bonded 158
Side-chain mesogens, doubly hydrogen-bonded 160
Size exclusion chromatography (SEC) 104
Small angle neutron scattering (SANS) 104
Spreading edge lithography 184

Star-[poly(styrene-*b*-2-vinylpyridine)] 55
Supramolecular polymer 1
Supramolecules, polymer-based
hydrogen-bonded 118
Surface grafting 101
Surface modification/patterning 183
Surface plasmon resonance (SPR) 186
Surfaces 1
–, hydrogen bonds 63
Switching protonic conductivity 142

Tape, antiparallel β-sheet 90
Templated self-assembly 184
Thermoreversibility, side chain bonding
146, 150
4-(Thio)pyridone 81
Thy-Au/Triaz-PS 193
Thymine/adenine 129
Thymine/triazine 123
Thy-PS:DAP-PS 190
Toluenesulfonic acid (TSA) 133
Trans-stilbazole ester 157
2,4,6-Triamino-triazine/barbiturate 46
Two length-scale structures 137

Uradiazole-hydrogen bonds 13
Urea 85
Ureidodezapterin 50
Ureidopyrimidinone (UPy) 82
– derivatives, quadruple hydrogen 124
Ureidopyrimidones 51, 62
Ureidotriazines 51

Vapor pressure osmometry (VPO) 105
Vesicles 193
Viscosimetry 104

Zinc-dodecylbenzenesolphonate 56

MIX
Papier aus verantwortungsvollen Quellen
Paper from responsible sources
FSC® C105338

If you have any concerns about our products,
you can contact us on
ProductSafety@springernature.com

In case Publisher is established outside the EU,
the EU authorized representative is:
Springer Nature Customer Service Center GmbH
Europaplatz 3, 69115 Heidelberg, Germany

Printed by Libri Plureos GmbH
in Hamburg, Germany